Climate Change

Climate Change
Seeing Beyond Tomorrow

Rollin Armer

ABOOKS
Alive Book Publishing

Additional copies may be ordered from the publisher for educational,
business, promotional or premium use.
For information, contact ALIVE Book Publishing at:
alivebookpublishing.com, or call (925) 837-7303.

Book Design by Alex Johnson

ISBN 13
978-1-63132-141-2

Library of Congress Control Number: 2021910275

Library of Congress Cataloging-in-Publication Data
is available upon request.

First Edition
(Previously published as "It's Up to Us")

Published in the United States of America by ALIVE Book Publishing
and ALIVE Publishing Group, imprints of Advanced Publishing LLC
3200 A Danville Blvd., Suite 204, Alamo, California 94507
alivebookpublishing.com

PRINTED IN THE UNITED STATES OF AMERICA

10 9 8 7 6 5 4 3 2 1

Foreword

For reasons that will become evident in this writing, I have been fascinated by the automobile. As I grew older, my interest in nature began to take root and grow, nourished mostly by a curiosity common to many people.

During the 1950s I noticed American automobiles becoming aggressively more powerful, at the same time more numerous. This resulted in a huge increase in visible smog which peaked around 1975. The Environmental Protection Agency was begun in 1970, and with what must have been considerable effort from automakers, has reduced visible smog to a tiny fraction of what it once was.

However, carbon dioxide from cars, a major source of global warming, is produced in direct proportion to how much gasoline is burned. Worldwide, the burning of coal is another huge source of carbon dioxide.

Methane gas is an excellent motor fuel produced by nature, but is a major contributor to global warming. Each molecule of methane contributes far more to global warming than a molecule of carbon dioxide. When used as a motor fuel, methane reduces the global warming otherwise caused by it. Methane is now being used in some commercial vehicles.

During the first fifteen years of my retirement, I divided my time between travelling the world with my wife and building and testing experimental automotive engines. They were designed with an eye toward better fuel economy, particularly at light load where they are working most of the time. My best engine managed a thirty percent improvement in fuel economy at light load. However, people in Detroit advised me that it would make too much smog and could not be improved. The U.S. patent for this engine is #6,672,270.

In this writing, I gather recent facts about our environment from the internet, specialized books, and a few people I am fortunate enough to know. As I gather input, I become aware of the swiftness with which our atmosphere is changing (not for the better) and feel a strong need to share with the reader what I find.

I hope that the telling of these findings will be of greater benefit than the saving of some fuel that my engine might have made.

I would like to dedicate this book to my wife Millie without whose help during the working days and companionship during the travel days, the impetus to write it might not have developed.

Chapter 1

An interest in machinery is born

I was born in Chicago October 28, 1929, the day before the stock market crash. My father, 26, had met my mother, 25, at U.C. Berkeley where he had earned a masters degree in physics and she, an english major, became a Phi Beta Kappa honoree.

My father's interest in physics was centered on sound, its production and transmission, and he was employed by Magnavox, Latin for 'Big Voice.' He designed one of the world's first loudspeakers, and applied for a patent on it in 1930. The patent (#1930576A) was granted in 1933.

Magnavox was centered at that time in Fort Wayne, Indiana, where we moved in early 1931. There I was to get a baby sister named Beret, born May 25, 1931. She turned out to be a very bright girl and always did well in school. I, on the other hand, and having been born with a blind left eye, was usually miserable in school. I often brought home notes saying "he's bright enough Mrs. Armer, he just won't try," or "lacks initiative" or "is constantly daydreaming." Also common was "is easily distracted."

I suspect I may have doubted the teacher's words and would accept for memory only that which I could see and hear. Some teachers may have noticed my doubt and been infuriated. If I felt the brass edge ruler on my knuckles, I would simply shut down, rejecting anything further she might say. Playing hooky was resorted to but had terrible results if my parents found out.

I believe my trouble with reading was genuinely due to my left eye not working. My mother, being an english major, really helped with spelling, grammar and, oddly enough, the multiplication tables. She played piano quite well and made beautiful renditions of the music of Zez Confrey, as well as many of the classics. Her knowledge of English and writing helped me considerably over the years.

I began to treasure what my right eye could see and made a record of it. There seemed to be inside my head a recording eye that, when interested, could record on a full color Rolodex everything I had ever seen. It was as though I said to myself "That left eye sees nothing, so select carefully what you see with the good eye, record it and treasure it. Keep it forever. It's all you have to gather knowledge and prepare for the future. There is nothing else." This would eventually help me tremendously as a mechanical designer.

I became fascinated by mechanisms, small and large, anything from a seventeen jewel watch to a steam locomotive, and everything in between- electric motors, door locks, locks of all manner, firearms, engines.

During grammar school I was very often daydreaming, possibly studying the octagonal pendulum clock with full intent to eventually understand how a small geared mechanism could store the energy of winding to turn the hands at a very uniform rate. I looked into the pencil sharpener to understand how two fluted cutters could turn backwards from the crank to shave the pencil to a cone.

The teacher might say, "Rollin, stop daydreaming and pay attention." Now, if there was a cement mixer running outside, the teacher didn't have a prayer. I would be entranced by the "pop-pop, pop, pop-pop-pop," with a "slush-slush" sound in between. Why didn't it have the same number of pops each time? After school, I would watch that mixer intently. It was made by Jaeger, the engine by Stover, with two flywheels, an oscillating magneto (a small electric generator that uses a permanent magnet to create the magnetic field; it was a self contained generator often used to provide ignition current on gasoline engines, and was used on party line telephones) by Wico, and a neat little glass drip feed oiler. A few years later, I would learn how the "hit and miss" speed governor worked. These terms and others may not be familiar, but will be expanded on later.

During the first half of the nineteen hundreds, the main power for portable earth moving equipment was steam. It was used on road rollers and shovels. I was in the third grade in 1938, and in our neighborhood, a steam shovel was digging a big rectangular hole for a new house foundation. The school day ended at three and the steam shovel operator left at five. This gave me two hours to study that thing in action and climb on it for study after five.

A typical small steam shovel was made of a bottom chassis with caterpillar tracks that could be driven independently forward or backward. In the center of the chassis was a large vertical spindle surrounded by a large bull gear. The main body rotated about that spindle and rode on a ring of rollers resting on the flat top of the gear. The body carried the boiler, three small reversing engines, the main boom pivot, two cable spools, the operator's seat and controls. About halfway out the boom rode a sturdy steel shaft on the end of which was a fork in which the bucket hung. I heard this shaft called the dipper stick. Where it passed through the boom was another reversing engine to drive the bucket toward or away from the boom. The bucket was essentially a steel cube maybe 30 inches on a side, open at the top with a drop open bottom. The front edge of the open top had a row of teeth to get a bite on the earth it would fill with.

The small reversing engines were an early form of servomotor wherein a load could be moved by a control far away. They were often used on steamships for such things as cargo lifting and lowering, anchor raising, and steering. They were activated by connection to a remote control valve with two pipes- one for live steam, the other for exhaust. The pipes traded function to make the engine run backward. The engine could be made self locking by driving a worm gear wherein the driven load cannot turn the engine. This way, if no steam is let in, the load is locked in place. Today, servomotors are usually hydraulic, and if the driven load is very far from the power source, they may be electric. Sometimes one can hear them whine when riding on an aircraft.

Near the top of the digging site was a path where dump trucks would wait to be filled. When the bucket was full, the shovel operator would raise it and swing it out over a truck and release the swing open bottom. A thirty inch cube of earth might be quite heavy, and the truck's springs would squawk.

The steam shovels were filthy with lubricating grease, a tar like fuel oil, and a coating of dust and grime. Upon coming home, I knew I mustn't come in the front door. When opening the back door, my mother might well say, "Wait, don't come across my clean floor- leave all your clothes there and get in the tub." Getting clean was one of the costs of learning.

I mentioned that my father had a masters degree in physics. This does not mean that he thought like a physicist of today for he obtained that degree in 1926. At that time, the study of physics applied more to the physical world familiar to many and less to subatomic particles, nuclear makeup and behavior, etc, requiring higher mathematics to even envision or represent what is happening.

My father enjoyed making things- musical instruments, jewelry boxes of ebony and rosewood, carving animals in ironwood, operation of a lathe, etc. He understood the transmission of the Model T Ford and how a steam locomotive works. His interest in photography led to the making of special cameras and photographic equipment for General Motors.

When I was seven, my father gave me a small wood lathe. He made a little bench for it and powered it with a ⅙ horsepower GE washing machine motor that he found in a hockshop for $2.50. He was very kind and gentle when showing me how to use tools, both hand and motor driven. I learned quickly and have always considered these things as an art. He was not patient, however. If I forgot something he said or went against his directions, he could make me wish I'd never gotten up that morning.

The positive often far outweighed the negative. If we were stopped at a railroad crossing waiting for a locomotive to start pulling a train, he would explain in some detail what each item was for and why. That white stuff coming out was NOT steam, it was water vapor. "You can't see steam," he would say.

I learned how the double acting cylinder worked, what the crosshead did, why the drive wheel with the main rod attached had a larger counterweight than the others. Eventually I would understand the Walschaerts valve gear and several others. He was especially informative when my mother and sister were not around- they might groan with boredom.

We lived in Detroit from late 1934 through mid-1939. Since my father worked for Magnavox designing radios and the auto industry had begun putting them in cars, he began to know people at the auto companies. When trade shows and auto plant tours would take place, he would take me to see beautiful machines. I saw a Pratt & Whitney radial aircraft engine that was all sectioned and cut away so that one could observe firsthand how it worked. It was being slowly turned by an electric motor and where each spark plug would normally be, a little red light would flash when the spark should occur. I started learning about valve timing, firing order, etc. There was a Miller 91 race car there, its little straight eight engine artfully sectioned to reveal the marvelous beauty of its inner workings.

My first lesson in thermodynamics (the study of the behavior of heat) happened when I was eight. My father and I were coming home after dark and had parked the car at the curb. Our house was not far from the Detroit City Airport in 1938. There was a Boeing 247 (predecessor of the redoubtable Douglas DC-3 as a civil transport plane) climbing from the airport. Both throttles were wide open and the exhaust stacks under the wing were both emitting a four inch diameter roaring yellow flame about four feet long.

We were walking toward the house and I grabbed my father's arm and yelled, "Hey, look- that airplane's on fire!"

"No it isn't," said my father calmly. "Come in the house and I'll explain it." He said, "The gasoline engine is a heat engine- it turns heat into work. But it can't turn all of the heat to work, because thirty percent gets away in the cooled cylinder surface and forty percent goes out the exhaust due to incomplete expansion."

Then he gave me an old textbook that he had used in college. The title was "Heat Power Engineering" by Barnard, Ellenwood and Herschfeld. It had a lot of pictures of heat engines of all sorts with explanations not only of the pictures but terms. This became instant input for my Rolodex and I began to understand terminology as well as function of various components and how they worked together. It began to dawn on me that some books had some really good things in them, especially if the explanations were accompanied by pictures. Then the words made sense. The pictures and drawings started my interest in mechanical drawing, not only as a means of representing something but proportioning the parts so that the assembly would function as a whole.

"Making it work on paper" became a new tool. Sometime later I found that college textbooks about engineering and machine design were far more valuable to me if printed before World War II. This was because they relied more on photographs, explanatory drawings, and pictures. More recent books tend to assume that the reader already knows what the subject looks like and dive into the mathematics right away. This brings in the problem that the reader may not have any notion of what a reasonable answer might be. Mistakes are not obvious.

One morning at breakfast in 1938 my father said, "You kids come home at noon today and we'll go out to the Ford factory and watch them build cars."

How good could things get? Miss some school and see the marvels of man's creativity unfold before our very eyes!

At that time the public was allowed far into the factory in segregated screened steel walkways beside and above major operations. These included huge rolling mills where orange hot steel slabs went through sequential rolls. At each roll stand, the slab would get thinner, wider, and go faster. The rolls, therefore, had to revolve a little faster as the slab progressed. If a roll stand was running too fast, it might smash the slab into the one ahead. If too slow, its slab might be smashed by the one following. Therefore, each stand was driven by a variable speed direct current motor of five thousand horsepower.

Somewhere I learned that to provide this huge amount of electric power were generators driven by Still engines. A Still engine (invented by William Joseph Still) is a steam jacketed internal combustion engine with double acting pistons as in a steam engine. Above the pistons, fuel is burned to expand, driving the crankshaft. The cylinders and exhaust system are surrounded by jackets where water is turned to high pressure steam. The steam expands underneath the hot double acting pistons to do further work. Still engines are very thermally efficient, but are never used in road vehicles because they are quite heavy and have a complex and slow throttle response requiring several adjustments.

The Still engine was run by burning carbon monoxide gas from the coke ovens. Coke ovens involve roasting raw coal to reduce it to pure carbon which goes into the steelmaking process. When the raw coal is heated, it gives off a large amount of carbon monoxide. This gas (besides being poisonous to breathe) is a combustible fuel of high heat value. Chemically it combines as $CO + O_2 \rightarrow CO_2 +$ heat. These two factors are easily adjustable to obtain best results.

The crankshaft forging process: The forging process squeezes out voids and impurities in the part giving good strength, fatigue life and uniformity. As I recall, there were two men on and two men off the press. The ones off stood before huge fans along one wall. All were wearing darkened face masks, leather arm protection, aprons and chaps.

The upper press die had three cavities in it and the lower die had a matching three. In the wall next to the press was a window with a trough sticking out. Rolling down the trough were the yellow hot cylindrical ingots. They were about six inches in diameter and sixteen inches long, weighing about 120 pounds. At the temperature of glowing deep yellow, they had the consistency of very heavy, thick bread dough.

One of the two men "on" stood at each side of the open press. Each had special tongs fashioned to grab either end of a yellow hot ingot from the trough. Laying it across the first of three sequential cavities in the bottom platen (or anvil), one man stepped on a control bar near the floor. This released the upper die, weighing perhaps seven hundred pounds and dropping about five feet. It smashed into the ingot with awesome force, sending sparks and bits of slag out horizontally.

Imagine closing a waffle iron suddenly after putting a little too much batter on to ensure the corners were filled. Now the forged piece has a thin flange or 'flash' around it needing a slight excess of metal in the ingot to ensure against voids. The second pair of cavities are to shear off the flash and prepare the ingot for the third strike. The third pair of cavities twisted the piece about the main bearing areas locating the "crank throws" in their proper position around the main bearings. Then the forgings were placed on a conveyer that would take them to further work as they cooled. It included rough machining, oil galley drilling, journal grinding, balancing, final measurement for assembly, etc. I imagine these tours very much increased a customer's desire to buy a Ford.

My enjoyment in being on such a tour rather than in school that afternoon defies description. No reading at all- everything I saw went onto the Rolodex so that now I can show it to you.

There were two other establishments in Detroit I was fortunate enough to visit: the first was the Ford museum where I saw all manner of early steam engines and automobiles with biographies of their originators, designers and inventors. The second was the restored laboratories of Thomas Edison at Dearborn, Michigan. The original laboratories and grounds were in Menlo Park, New Jersey. In 1938 there was a white haired gentleman showing tourists through the restored laboratories at Dearborn. He was Francis Jehl and had been an assistant to Mr. Edison for some years.

Museums are of exceptional value to me because I can see the results of man's creativity firsthand and record them on the Rolodex. Merely reading about them is, for me, a much less productive use of time. If the writing is highly descriptive, it helps.

I should comment on two other museums that, to me, were so fascinating that I nearly wore out a pair of shoes in each one. One was the Deutsches Museum in Munich, Germany, where I learned much about the opposed piston Diesel aircraft engine of Hugo Junkers, the electric dynamo designs of Werner Von Siemens, and much other early German technology. The other was the Field Museum of Science and Industry in Chicago, Illinois. They had a steam locomotive there with animated pressure-volume diagrams in colored lights, and a captured German submarine, the U-505. There was a fascinating account of how it was taken and brought to Lake Michigan via the St. Lawrence River and waterways between the Great Lakes.

Chapter 2

Moving to Davis, California

By 1939, my parents were growing weary of life in Detroit. The climate and soot laden air were conducive to whooping cough. Beret and I both nearly died from scarlet fever and many of our schoolmates missed a lot of school. I, of course, didn't mind missing school, but being sick interfered with my learning.

All my parents' relatives lived in California and would welcome us. So, my father secured a job with Agricultural Engineering at the University of California, Davis campus. In the summer of 1939, my mother (who was pregnant with my younger sister Elinor) and my sister Beret and I came west on a steam train called 'The Challenger.' My mother's mother lived in Berkeley and the three that came on the train lived with her while my father found a house for us in Davis.

During September and October of 1939, Beret and I attended Emerson School in Berkeley. It seemed to me much friendlier than Halley in Detroit. My teacher was Myrna Gifford and she was so reasonable that I got along well.

In 1939 the East Bay, including Berkeley and Oakland, had a very good interconnecting railway system of which the 'Key System' was a part, in which streetcars could be taken almost anywhere one wanted to go. The population at that time was less than a tenth of what it is now, and the number of automobiles must have been less than five percent of present numbers. Berkeley seemed a beautiful city and I felt that the moderate population density was one reason.

In Detroit at night there was a continual din from the large dense surrounding city. Berkeley had none of that, and while lying in bed, one could hear an occasional streetcar stop, pick up a fare and growl away with some slight hiss of the trolley. The kids at school were more open and adventurous than those in Detroit.

My father was at that time working in Davis and staying at a hotel there while he found a place to live for us. What we settled into was a lower flat near the downtown area. Davis, like Berkeley, was quiet at night. The only interruption, since our home was three blocks from a railway switchyard, was the occasional engine whose tender was taking on water. When the tender was full, the iron door would close with a clang and the engine would start, first with a clack-clack-clack as the car couplers took hold. Then, if the side rods were loose, a chunka-clanka-chunka-clanka off into the night.

My fifth grade teacher there was far less sweet than Miss Gifford had been. She had a simple way of regarding boys who had trouble paying attention: they were 'bad' and their likelihood of doing anything decent was always in doubt. Then there was the class bully who was somewhat lazy and didn't like to expend needless energy. Since I was small and thin, he would delight in showing me how far back he could bend my wrist and elbow without breaking anything. Many of the other kids were friendly and in the evenings we would play kick the can or darebase or red rover. Here there was a pretty good life outside of school. My sixth grade experience was a lot better. We had a man teacher who was sympathetic to my reading problem and believed I was a nice and worthwhile person anyway.

My second vivid awakening to heat loss in man's engines occurred when I was eleven. It was a weekend morning and I walked down to the switchyard to see what I could. I was standing next to a switch engine. It was on a siding that connected to a mainline that went to Dixon and points west. It had steam up and I was staring at it overwhelmed with curiosity and wonderment. The fire made a low rumble, some steel parts were sweating oily drops of hot water and the air brake pump made an occasional "toong toong."

After a moment, the driver and fireman climbed down and walked over to a coffee shop a half block away. A couple of minutes later that engine began to move forward. I ran over to the coffee shop and yanked open the door. There they sat and I yelled "Hey- your engine's going down the track." They jumped up and ran after their engine. The mindless thing was doing about two miles an hour when they grabbed it. They climbed on and brought it back. I was still standing there and rather than thanking me, they said, "Hey kid- climb up on this thing- we'll show you how it works."

In the next twenty minutes my learning curve must have gone straight up. The fireman said, "Here, let me show you the fire." He stepped on an air valve button on the floor and the fire door swung open.

When I saw the inferno in there, my first impulse was to yell, "Let's get out of here." But the other two didn't seem frightened and I didn't want to look chicken, so I kept quiet.

Years later when I understood the modest thermal efficiency of the steam locomotive, I remembered the sight of that blaze. I realized that rate of heat delivery was required just to keep steam pressure up. In other words the engine leaked heat at that rate. If you wanted to pull a train, you'd have to make it hotter than that.

My parents had been planning a home for us and had engaged a builder in 1941 who was able to secure all material before the war effort made such things impossible. The house was nearly complete when Pearl Harbor was bombed and we moved into it later that December. The house was built on the west side near the north end of Oak Avenue. This was a new suburb of Davis just west of and parallel to College Park. It was a north-south street then a quarter mile long, the south end connecting with the main highway through Davis. The north end was fenced off to fields often planted with sugar beets or barley.

Just south of the highway was the Davis campus of the University of California. Of the campus' many buildings were some with lecture halls and laboratories, and several dormitories. The new Chemistry building was being completed and there was a complex called Dairy Industries along with several barns where experimental animals were kept. All this was within easy bicycle distance from my home and I became quickly interested in the Agricultural Engineering building. There people designed and fabricated new and experimental farm machinery, including peach pitters, almond huskers, walnut shellers, etc.

My father had an office there and was assigned the design and fabrication of a sugar beet harvester. There was a machine and fabrication shop there where the experimental machines were built.

One of my first jobs, since I had learned some drafting, was to make inked drawings of machines and parts on linen for blueprinting. This job was during Christmas vacation and I received 25 cents per hour. But the main point of interest for me in that building was a course for Ag students on the operation, maintenance, and overhaul of farm tractors and various engines used on farms. There was a very large lab room there with a row of assorted tractors including track type and twenty some stationary engines including four Diesels. There were two long tables, one with various kinds of magnetos, some sectioned and another with many types and kinds of carburetor. By studying these things first hand, true understanding could occur.

The course was led by a very kind and knowledgeable professor named Ben Moses. He was also the scoutmaster for a while and one of the best we had. He held a class there every Saturday morning, and I was always there, entranced by the fascinating machines. One Saturday, Dr. Moses came to me after class and said, "Rollin, there is a Novo engine over there that the students didn't get to. It's completely apart. Why don't you come back after lunch and we'll put it together and make it run." I was so excited, I didn't eat much lunch and rode my bike back there before one o'clock.

The engine was a vertical hopper cooled type with two flywheels. It was hit and miss speed governed and had a bore of about three and a half inches and a stroke of five, denoting the piston diameter and how far it moves up and down in the cylinder.

Dr. Moses showed me how to cut out a cylinder base gasket with a small ball peen hammer, how to connect four large dry cells in series to get six volts for the ignition coil, and how to mesh the timing gears so that the exhaust valve opened and closed at the right time relative to the other parts. In a couple of hours, it was ready to try. Ben showed me how to fit the crank so that if the engine kicked back, my hand would be free and not get hurt. It started quickly and after a few chuffs, came up to its governed speed of 500 RPM.

I was elated. Not only had I made it run, but I had noticed that some older people really know a lot about the things that fascinate me. Therefore those things must be of value somewhere. Since this happened to me in a school, maybe school isn't all bad. At least, as I grow up, I should take a serious look at it.

During these times, I got through the seventh and eighth grades, picking up most of my knowledge outside of school. I managed to get grades only good enough that I wouldn't have to repeat anything.

My seventh grade teacher was brand new at teaching, twenty three years old, and had a nice combination of black hair and blue eyes. I fancied that I loved her and sometimes tried to get her attention in ways that often had terrible results. Fortunately, my teachers from then on through high school were spared the effects of any love I might have had for them.

I will have to be careful not to portray these years as worse than they actually were since my memory of them was of failure, embarrassment, and being often pushed around by other small people.

When I entered high school, I was small at five feet three inches and 115 pounds. Four years later, upon graduating, I was the same height and weight. I had an odd sense of dignity because of my Rolodex which gave me a know-how about things, not people. Nobody could take that away, so I revelled in it. Unfortunately it consumed nearly all of my attention.

My sister Beret, a year behind me in school and a year and a half younger in years, was the tallest girl in her class. She always did well in school. With her early maturity, good looks, and reddish brown hair, she was the envy of some of the other girls. Our interests were so far apart that we were never very close.

My scholastic progress in the time was a disaster. I often felt afterward that I could have done anything else for those four years and been the better for it. Being blind in the left eye and poorly coordinated, I couldn't catch, pitch, or hit a baseball. Football was a disaster and I remember being knocked out twice. The coach was not pleased but sympathetic and gave me fair grades for trying.

My first car was a 1926 Star. It had a 2.5 liter four cylinder Continental engine. The steering and rear wheel brakes were terrible. There were no front wheel brakes. In late 1943 it resided in a barn and some claimed that it would run if it had a battery. I bought it for $15. I was in possession of an old Willard battery that I kept charged with a junkyard washing machine motor driving a junkyard Autolite 6 volt generator. One Saturday I and an adventurous neighbor boy named John Taylor tied a small wagon carrying the battery to one of the bikes and rode to where the barn was, containing the prize. The original fuel system was gone, and in its place was an ancient dash mounted air pump to pressurize the gas tank. There were no doors or top and the air system leaked so that John would have to stand on the left running board and pump continuously while I drove only in high gear since the clutch didn't work either.

The car had no starter and the generator was burned out, so to start, we had first to pick the right pair of wires to twist together for ignition, then after pumping air for the fuel system, we put it in high gear and pushed. It went. Fortunately there were no stop signs between the barn and my home because after we left the barn, we would not have the strong farm boy that I bought it from to help push. At the end of the country road, we turned left onto the highway. Then, with John manning the air pump, I brought it up to about thirty miles an hour, and was overjoyed. Listening to the little engine chortling happily, I began to feel a kind of kinship with my new possession.

Later, for an English class in school, I wrote a story of that experience and got an A on it. Apparently I could write easily about real experiences because they were on the Rolodex. However, writing about things that had no pictures, like dates in history, algebraic relationships, sequences of button pushings on a computer to get some result was and still is virtually impossible for me. I can't get a handle on it.

As high school progressed, I became known for earning rock bottom grades and getting anyone's jalopy to run. This was during the war years and any car more modern than a Model A Ford was usually beyond the means of a high school student. So, hiding in junk yards were a number of old cars, some that, with clever manipulations, might be brought to life. Driving one of these down a county road to avoid the police, burning cleaning solvent to circumvent the need for rationed gasoline, without a muffler to appreciate the engine's release from dormancy, was a thrill that rivaled all others in those innocent days. Before long however, the bawling out by police and the paying of fines etched away at the social freedom and turned out, with luck, more acceptable citizens. Those days were full of memories, some very nice except for those of school.

Summers on the other hand, were usually interesting because I was always employed at something, interested in making it go right, and had no time to play movies in my head. The summer after my freshman year was filled with odd jobs repairing household appliances and refurbishing discarded items for further use.

After my sophomore year I worked on the dairy farm where my first car had come from. As I recall, the cows were Holsteins and one of the feeds was alfalfa, all of which was raised there. That job consisted of tractor and buckrake driving for the handling of hay- mowing, side delivery raking, gathering the windrows into haystacks and, with the buckrake, carrying the stacks to a barn where they were lifted into the barn loft. The people there were very nice and my pay was 50 cents per hour.

By the time of my junior year, I was gaining a reputation of being able to fix things including cars and trucks. One of my school friends knew of an automotive engine rebuilding shop in West Sacramento where he had had much work done. He suggested I try there for a summer job. I went there and talked to the owner and was accepted. The shop had a crankshaft grinder, cylinder reboring and honing machines, bearing and rebabbiting equipment as well as valve grinding machines.

Any device that would be required for the complete rebuilding of automotive engines was there. The year was 1946 and Detroit had just begun to manufacture post war cars. Therefore there was a great deal of rebuilding of prewar cars going on. There were three ex-servicemen working there. One had a wooden leg and they all seemed happy to be back alive and to be employed. They showed me a great deal about refitting and proper assembly of car engines. This job paid 65 cents per hour.

Like servicemen anywhere, the three there turned the air blue with profanity. I picked up on it, thinking it was necessary to make the job go right. I learned a great deal that summer including how to watch my mouth in polite company.

The job I got the summer after I graduated from high school was a real eye opener. It was with the Animal Science department of U.C. Davis and I was to work for Dr. Max Kleiber directly. The Animal Science building in 1947 was located on the southwest corner of the quad, and it had a small machine shop in the basement. Dr. Kleiber had been designing an apparatus he called a 'respirometer' for dairy cows. Its purpose was to measure the amount of oxygen inhaled and carbon dioxide exhaled by a cow in producing some measured quantity of milk. My job was to build the machine. We had access to a good glass shop and could get nice custom work done on any glass requirements like large retorts, flasks, and condensers.

To measure the volume and frequency of breaths, we made an open bottom closed top cylinder about sixteen inches in diameter and a foot high. It was made of thin monel, a nickel alloy, for its corrosion resistance. It hung on a half inch diameter steel rod going up to a chain. The chain went over a large pulley down to a balance weight. This inverted 'bucket' was free to move up and down. It was partly submerged in an open top vessel having a 1 ½ inch diameter tube protruding up from the bottom about eleven inches. This bottom pipe connected to large free flowing check valves (that came from war surplus gas masks) such that the incoming air went through glasswork measuring oxygen concentration.

The exhaled air went through other glasswork showing the difference in carbon dioxide concentration from the inlet air. The increase in CO_2 was proportional to the decrease in O_2. From the bottom fitting on the outer tank came a flexible rubber tube about 1 ¼ inch diameter. This tube had a metal fitting with a very soft, delicate rubber collar. The collar was inflatable by a little squeeze bulb outside. It was made from a condom and served perfectly. The very soft collar on the tube was to connect to the cow's trachea which is lined with mucus membrane. If this delicate membrane is breached in any way, infection is bound to occur.

Now, as the cow breathes, the inner water sealed 'bucket' moves up and down. On the vertical shaft that carries the bucket is a horizontally held pencil pressing gently on a strip of paper. The paper is wrapped around a vertical axis cylinder about a foot high and six inches in diameter. The paper cylinder is rotated by a clockwork motor at exactly one revolution per minute. After one minute of cow breathing, the paper is taken off and shows an accurate curve. The curve shows very closely the volume of each breath and how many there were per minute. The measurement of gas concentrations was made chemically and I never knew how it worked.

Here, another lesson blossomed in my head: "Everybody knows something and nobody knows everything." Max Kleiber knew a great deal.

That fall, since I came from a college-educated family, I was expected to continue into college. This expectation filled me with dread and I knew in my heart that trouble was afoot. My high school performance had been so poor that I could only enroll in a junior college and then try to get C grades in high school-level courses that had previously been below that. So I enrolled in Sacramento Junior College in the fall semester and took again those courses that had stumped me before. I did so poorly that I walked out, convinced that I must start to work then because college was not an option. That spring and summer, I held jobs that were like the summer jobs I had had before.

During my semester at Sacramento J.C. I had noticed that they had a Technical Institute of Aeronautics. That fall I looked into it and found that I was eligible to try their route, which I did. I began to feel at home there and took a course in General

Aeronautics as well as getting practical experience working on aircraft and overhauling their engines. For the first time in my life I was getting A's and B's. By the spring of 1951, I came out of there with an aircraft mechanics license and an Associate of Arts degree.

Another very welcome thing occurred during that enlightening school experience: I grew to over six feet and normal weight.

The Army

When I was twenty one and no longer in college, I was about to be drafted into the Army. I had heard terrible stories about it and so tried to enlist in the Marines, the Navy and the Airforce, but none of them would have me because of my blind left eye.

There was a story about the Army wherein they felt your arm. If it was warm, you were in. This turned out to be true. After taking basic training at Aberdeen Proving Grounds in Maryland, I was able to talk the classification and assignment team into sending me to San Marcos, Texas. There I would learn assembly and maintenance of liaison aircraft, both fixed wing and helicopter. This was right down my alley and I got through it well. After that training, the class of about twenty I was in were seated in a room where we would be assigned our final place of duty.

Since it was the fall of 1951, there was a deadly war going on in Korea. An officer in the assignment room stood up and said, "We need one volunteer to go to Alaska." My hand shot up, not so much because I would avoid Korea, but that here was a chance to control my own destiny, a beautiful thing seldom found in the military.

In late December of 1951, I flew to a point of embarkation near Seattle, Washington, called Fort Lawton. It was very cold and the air was laden with coal smoke, far enough north that the days were short and gloomy. Fort Lawton is no longer there.

After two weeks there, I and about forty other servicemen, some going to the far east, boarded an old and slow steam driven troopship named the USS James O'Hara. It must have had a round bottom for it wallowed and rolled in heavy sea. And when the stern rose, the single screw would come partly out of the water and shake the aft quarters enough that the seasick passengers had trouble holding onto the sinks and toilets they were throwing up into. Crew members were going around with steam hoses and mops to maintain a semblance of order. It is hard to imagine a more miserable means of going somewhere. From reading some history of early sailing vessels with disease, darkness and death, one can believe it could have been worse.

After a week of the foregoing, the ship came into Cook Inlet where the sea was kinder and tied up at Anchorage. I and some of the others boarded a train to Fairbanks. The train ride was nice- there was no motion sickness. From Fairbanks we boarded a bus that took a white snowy road 106 miles southeast to our final destination: Big Delta.

Today, Big Delta is a small city. When I arrived there at the beginning of 1952, it was a small Army base. It had two divisions: the Army Arctic Indoctrination School and the Arctic Test Branch.

I was attached to AAIS and our function was to show field officers from the U.S. how to get along in freezing weather; travelling by truck, tracked vehicles, large sleds, aircraft, motor boat, even dogsled as well as camping for days at 40 degrees below zero. The Arctic Test Branch had the function of testing clothing, vehicles, heaters, tents, parachutes, lubricants, starting fluids and anything that might be necessary for survival under similar conditions.

There was a large aircraft hangar parallel to an east-west runway which was about two miles long. Large cargo aircraft like C-119's could land there. The hangar was two stories high with a sliding door and apron at each end. There was an aircraft radio tower and a fire station nearby.

The base had the NCO (Non Commissioned Officers) club, PX (Post Exchange), Post Office, Mess Hall, and barracks buildings. The barracks were in the form of an H. They had steam heat and were quite comfortable. Electric power came from a nearby power plant with three large Fairbanks Morse two stroke Diesels. They ran at 300 RPM and lent a constant gentle boom to the atmosphere. It was somehow reassuring in winter when the sun rose at 11:30 a.m. and set at 1 p.m.

AAIS also had a waterways section with a number of Tanana boats. These were a long flat bottom boat of very shallow draft that could negotiate the wide shallow Tanana and Delta rivers.

For leisure time there was a crafts shop and a library. I seized on the library with the notion that I could teach myself algebra. Finding books on it there, I tried for several evenings, but got nowhere. The crafts shop, however, was very nice. I and another fellow made a small steam engine out of old shell cases. It burned canned heat and had a whistle that people didn't like.

Also for leisure, two or three of us would get a three day pass and take a Tanana boat with an outboard motor on a fishing and camping trip. The river had some number of grayling in it. They were usually twelve to sixteen inches long, related to trout, with a long colorful dorsal fin. Very good to eat.

One very important building to us was the motor pool. A large sheet iron building housing some "Deuce and a half" trucks, M-29C weasels, Caterpillar D-8 bulldozers, etc. A Deuce and a half truck is a six wheel, all wheel drive truck rated at two and a half tons. I have seen one carry nine tons without trouble. The springs are so stiff that, if one has to ride in the back any distance, one would swear they must have square wheels.

The weasel was a tracked vehicle that normally was pretty good on snow, but the tracks, instead of being steel links, were rubber belts with cleats bolted on. If the weasel was running in snow on the side of a hill, sometimes a track would run off the drive sprocket and idlers. There was an iron-clad guarantee it would be the downhill track. This way it would get behind the sprocket and be most difficult to get back on. It usually took some cut tree branches and just the right color of profanity to get running again. The weasel was basically a flat-bottomed hull with slanted ends something like a landing craft. It had an arctic cab fully enclosed in windows with a powerful heater. It had a winch capstan in front to pull itself or other things out of holes but we never had to use that.

Both the AAIS and the ATB had an air section in the hangar. AAIS had an L-17 and an L-5 aircraft. The L prefix means liaison, usually a single engine small plane with a very good visibility and good low speed flight characteristics. Their design purpose is to preceed armed units into a battle area and radio back what is going on. The ATB had two Cessna L-19s, a Bell H-13 helicopter and in 1952, got a DeHavilland Beaver. The L-17 was the military version of the Ryan Navion, a four place low wing craft with a sliding canopy. The L-5 was the military version of the Stinson flying station wagon, high wing tandem seating with a side door aft to carry a litter patient. It had very good visibility and good low speed flight characteristics. Those were helped by wing slots ahead of the ailerons, and the ailerons could be dropped considerably, maintaining their function at low speeds.

I was to be crew chief on this plane. According to its log, it was built in 1943 and had several thousand hours of flying time. It had a six cylinder Lycoming engine of 180 horsepower and an Aeromatic variable pitch propeller. Normally it was used to check out weather and snow conditions in areas where officers would be training for camping and survival. Once we used it to carry out a litter patient. He had gotten a twig stuck in his eye. The L-5 was able to land nearby and fly him to a hospital in Fairbanks.

In the spring of 1952, the Army wanted to study the logistics of having the L-5 on floats. This way it could land on lakes, bringing people and supplies to Alaska's lakes as well as looking at the progress of the officers-in-training. Landing on snow with a plane on wheels can be treacherous because the snow may pile up before the wheels, causing the plane to nose over. We did this once with the L-5 and damaged the propeller. If the plane is equipped with skis, it is okay for smooth snow or ice but nothing else. So it was decided that the L-5 would be on floats (or pontoons) and be based at Bolio Lake.

Since I was the mechanic on the plane, I would be camped at Bolio Lake with it. Bolio Lake was about three miles long and a mile wide, running north-south and was twenty miles south of Big Delta. I was furnished with a Jamesway Hut, about a half truckload of lumber, mostly 2x4's, a great many tools and my M-1 thirty caliber rifle. I had ample clothing and some boxes of C-7 rations to eat. From a friend at the base, I borrowed a .22 rifle with a scope sight and a small fishing rod.

Of wild creatures to be wary of, the Alaska brown bear was the worst. Clearing away the myths, one of these could stand nine feet tall and weigh up to 1700 pounds. There was a steadfast rule about them: don't ever shoot at such a bear unless it is trying to kill you- because it you do, it will. Now, however, if one of them is trying to kill you, and you have your M-1 rifle, try to have gotten some .30-06 cartridges with 220 grain bullets. They have a blunt lead nose and will smash rather than penetrate, giving a sledge hammer-like blow.

The military .30-06 weighs less and is copper jacketed with a sharp point. It is designed to wound an enemy rather than kill him because a wounded person takes two others to help him. While they are doing that, they are not shooting back.

I tested this ammunition as follows: I sawed off fifteen pieces of 2x4s about 10 inches long and nailed them together in a stack 22 inches thick. Taking the M-1 and loading a single round of military ammunition into it, fired into the stack. The stack jumped a little but nothing came out. Then I loaded a round with the heavy soft nose bullet and fired again into the stack. A puff of splinters blew out where the third 2x4 had been. The standard bullet made a clean round hole through twelve pieces and stopped in the thirteenth. This gave me confidence that I had reasonable protection against huge bears and during my six weeks there, I never saw one. I really would not want to hurt such a respectable creature, and if I did confront one, would I always have the rifle at the ready? Probably not.

A Jamesway hut is an arctic shelter in the form of a half cylinder lying on its side. It is about fifteen feet wide and the length is variable depending on how many sections one wants to add. Mine was a single section about fifteen feet long. The floor is made of flat plywood boxes about four inches high and three feet square, linked together. The top is a half circular arch made of two layers of canvas snapped onto bows that are fastened to the floor. One end is blanked off by a canvas wall, the other has a hooded doorway with outer and inner doors. There were two screened windows with blinds on either end. The top arches and floor panels could be linked together to make a shelter as long as wanted.

I took some lumber and made a cook shack a little distance from the door. I did not want any food scraps near the door of the hut. Going up the hill about fifty yards east of the hut I dug a slit trench for my toilet and a garbage pit. This way any interesting smells would not attract large animals to the hut. I had several cartons of C-7 rations in the hut but did not think they would smell because all food was packed in sealed cans or aluminum foil. Inside the hut, to sleep, I had a hammock with an arctic sleeping bag and a mosquito net. With some of the lumber, I made a table and a chair.

The L-5 aircraft was at Big Delta and sitting on wheels. We had two sets of pontoons (used and not without leaks) that we trucked into the base from Ladd Field in Fairbanks. To put the plane on the lake, we proceeded as follows: take off the wings and struts and the landing gear. Put the plane on one truck, the wings and struts on another and the pontoons on a third. Drive the trucks to the northeast corner of Bolio Lake where it is fairly flat and sandy. The pontoons were made as a single unit, held apart by two large streamlined aluminum tubes and X bracing of streamlined rods with turnbuckles. They bolted onto bosses on the lower corners of the fuselage frame where the wing struts and former landing gear legs had been fastened.

Before bringing the plane to the lake, we had trucked out a Caterpillar D-8 bulldozer. With it came a very good driver named Everitt Wayne Sparks. We carefully drove the dozer into the water to see if it might sink or whether the water might get into the engine or gearcases. It didn't seem to, so we dozed out two slots in the shore- one to park the plane and the other to tie up a boat if we should need it.

A plane on floats is very hard to manage in a light wind if it is not tied up and the engine not running. The large rudder will turn it directly into the wind, then it will back up until something stops it. To assemble the plane, we put the floats in shallow water, resting gently on the sand bottom. Then we bolted the fuselage to the float assembly. As it got heavier we lifted and pushed a little farther out, bolting on first one wing and its strut, then the other. Then, making all the internal connections, fuel lines, control cables, electric wiring, fairing strips, we manhandled it over to the slip and tied it up.

On my first leisure evening there, I went into the lake to take a bath. The bottom was all sand and gently sloping with no surprises, so I swam a bit. After a while I came out and discovered three leeches on my legs. They looked very evil and as I pulled them off, made a note to keep my baths shallow and short. Then, sitting on a rock near the water and not making any noise, I heard a little thump-thump-thump. Soon I heard it again, and saw a snowshoe rabbit. Presently I saw another and began to wonder how close I could come to being self sustaining as far as food and shelter were concerned. I don't know just why, but it would make me independent of social demands. I imagine this goes on in the minds of many people who would live this way.

So I began to study my environment. I found in the meat department rabbits, many tame marmots (an Alaskan type of woodchuck) and ptarmigan (an Alaskan grouse something like a pheasant that turns white in winter). I tried fishing in the lake, but only caught some very small bullheads. This lake froze to a depth of about four feet in winter so I'm not sure any decent fish survive there. There were wild blueberries and strawberries there. Probably, due to the lack of people, the animals there were quite tame. The marmots would come to the door and chatter at me to give them food. They liked crackers, maybe for the salt. I saw one ptarmigan close by and it was so pretty, I chose not to shoot it. I developed a motto- "don't shoot any animal that you don't want to eat." Shortly after I made camp, I shot a rabbit. It tasted so good that, while I was there, I ate thirteen of them.

Wanting to explore the lake some, I decided to make a raft. I took four empty 55 gallon fuel barrels and, with some two by sixes, made a frame that would lie on top of them while encasing two pair. Each pair was end to end and the two pair were side by side. I tied them up under the frame with rope. With sticks through the knots, I twisted the ropes tight and left the sticks in.

A raft is a very clumsy thing to paddle, so I got a two horsepower Briggs and Stratton engine from a Herman Nelson heater that was very common at Big Delta. I cut a piece of two by four and, with the saw and a hunting knife, fashioned an air propeller. The engine had a flange on the output shaft so I bolted the propeller on it. I balanced the prop as best I could by marking one blade and running it- then shaving a blade with the knife and trying it. After a few tries, it ran smoothly enough.

I pushed the raft into the water to see which way it pushed more easily: with the barrels going end forward or side forward.

It turned out that forward was easier so I bolted the engine on what was then the back. On the water it was very stable, but slow. It made about a slow walk for progress. Most of the shore was covered with reeds, weeds and small trees, not sand. Some red legged shore birds lived there, but I did not see much else.

According to today's maps, there are some roads, buildings and activity especially at the south end of the lake. In 1952, it was just me and the animals. Occasionally some officers would come out and fly the plane, planning training maneuvers for groups from the States.

Two weeks after I had established my camp, about five of my friends from Big Delta came out in a truck to bring fuel barrels, my mail, and other supplies. They all brought their trusty M-1 rifles for some conjured up reason and really shot the place up. Great sport. A lot of rocks and things that might have looked like an animal got killed. It was fully two weeks after that before I saw any signs of life. It established in me a notion that people and nature are not always compatible.

One night I was awakened by a crunch-crunch sound. I was somewhat alarmed and took the M-1 rifle quietly outside and walked around. Finding nothing, I went back to bed. There was the crunching again. This time I took the .22 rifle outside and waited. The crunching started again and it turned out to be a porcupine eating the corner of one of the hut's floor panels. It turns out that porcupines are not easily alarmed and don't go away when you ask them to. I never saw such a "so what" animal. Not wanting more of my floor eaten, I shot this creature in the head with the .22. It simply dropped with no sign of stress and was dead. I picked it up by one foot with a pair of pliers and buried it near the garbage pit. It must have weighed twenty five pounds. Later, back at Big Delta, I was told that one should never kill a porcupine because they are the only animals a starving person can catch and eat. Glad I wasn't starving.

Mentioned earlier was the fact that the pontoons on the plane leaked. The pair we chose to put on the plane leaked the least we were told by the people at Ladd Field and the left one hardly leaked at all. The floats were very nicely made of riveted aluminum. Each one had seven compartments separated by bulkheads, and each compartment had a handhole and cover on top. When we first assembled the plane on the water and tied it up in the slip, we didn't take it out for a couple of days. On the morning of the second day, the right pontoon had sunk somewhat. One of our two pilots suggested he could fly it long enough to let the water leak back out. With just him in it, the plane was light and he managed to get it off O.K. Indeed, the water started running out in the air, but not very fast. By the feel of the plane, he radioed that he couldn't stay up long enough to empty it, and came back. We located the compartment that had the most water in it. After trying other methods of pumping, we finally settled on a fuel barrel transfer pump. After that, and before every flight, we would pump out the right pontoon. We used the plane for planning more trips of officers from the States to get camping experience.

After I had been there six weeks, we dismantled the plane and brought it back to Big Delta where we reassembled it on wheels. The Army took back their hut. A few months later we flew the L-5 over the lake as it was beginning to freeze. The camp was totally gone, but the trusty raft was still there. I hoped someone might make use of it.

The Bolio Lake experience left me with a new perspective. A piece of the world with people and all that they bring with them as opposed to that same little piece without the people and just the numerous other inhabitants. The other inhabitants were self-limiting in number so that the prevailing resources would support them. The people and what they bring have a much greater impact on the environment that far outstrips their basic needs.

Since that experience, I have never been certain that the

people had sufficient understanding of their environment to survive indefinitely. Today, sixty five years since that time, I am still uncertain. Perhaps this uncertainty is what propels me to show the reader the beauty of what we have and to encourage the effort required to keep it that way.

After the Bolio Lake experience, we got another mechanic for the AAIS air section. His name was Jim Wright. He came from Texas and he had gone through the same light aircraft schools in San Marcos that I had. He turned out to be a nice person and a very good mechanic. Not long after Jim Wright arrived, our air section got a brand new Bell H-13 helicopter. Well, brand new does not mean assembled. What we got was a big planeload of boxes. Jim and I accepted the challenge and determined to make it the best flying helicopter in the Army. We had it assembled, carefully made and measured all the ground adjustments within a week.

Our pilot for that craft was a Captain Foy F. Ketchersid. He was already with AAIS and had rotary wing experience. Jim and I had moved the assembled machine out onto the west apron of the hangar. The day was clear and sunny.

Captain Ketchersid said, "Is she ready to fly?"

I said, "Yes sir."

He said, "Get in." I got in and fastened my belt. He did likewise. He surveyed the instruments, adjusted the altimeter, turned the ignition to "both" and hit the starter. The Franklin engine started immediately and came up to 600 RPM. The main rotor slowly followed and began a gentle "whish whish." The oil pressure was fully up in a couple of seconds. He left the throttle there for a few minutes then ran it up to about 1200, holding it there for about fifteen minutes. The main rotor turned one tenth of engine speed and there were two concentric needles on the tachometer. When the closer needle was just over the back one, it meant the clutch was fully engaged and all was fine. The engine temperature was now normal and the captain took us up about two thousand feet and flew west about twenty miles. There he climbed vertically to about nine thousand feet where he held it stationary.

He then yelled at me, "You have the collective pitch stops adjusted?"

I yelled, "Yes sir." Thereupon he pressed the lock button on the collective pitch lever and pushed it down. The engine came down to an idle, the rotor went into autorotation and we descended smoothly. At about five thousand feet he brought the pitch stick back to normal and set the lock. The engine went back to three thousand turns a minute and the machine flew normally.

I used to smoke in those days, and feeling relieved that my adjustments were made properly, I lit a cigarette. He yelled, "Corporal, give me that cigarette and take this thing for a while." This was a tacit way of saying 'thank you for your work' as well as an extension of trust.

With a combination of pride and uncertainty, I took the controls. For about twenty minutes, that thing was all over the sky- up, down, right slip, left slip, backward, forward, etc. Then, just as when I first learned to ride a bicycle, I got the feel of it. I stopped overcontrolling, got used to its response time, and it started to behave like a tamed animal.

The captain yelled, "That's pretty good, let's go home." He made no move to take it back so I flew it back to Big Delta. When we got to the hangar, the two story-high door was open. I hovered in front of it for a moment, and he yelled, "go ahead, take it in." I did, set it down, and letting the engine idle a minute to cool the valves, cut the switches. It was quiet except for the 'whish whish' of the coasting rotor. We looked at each other and felt a mutual respect.

Northway

Later that fall, a number of us from AAIS were to camp at Northway, about a hundred miles southeast of Big Delta. The purpose was to try out some experimental shelters. We had three, and when assembled, they looked like large double walled teepees. There were several fiberglass tubular poles to support them. The temperature there hung steadily around -40 degrees. Daylight was short and we were to prepare our rations on little canned heat stoves. Light was provided by Coleman lanterns. Between our breathing and the stoves, fresh air was needed. It came in between the canopy and the ground and went out at the top. We had some trouble letting in enough air to breathe and staying warm at the same time. I believe, after our experience with these shelters, that word was sent back to their maker that further development was needed.

There was a large major highway going right by Northway, and we had landed the L-5 on it when we arrived for our cold shelter experiment. There was a small hangar there to accommodate the aircraft. As I recall, the hangar was open front and had no heat.

On the day we were to leave, the temperature was still -40 degrees. It was about 11 a.m. and the weather was clear. The truck drivers managed to get their vehicles loaded and started. They were warming up in a cloud of ice fog. It occurs in very cold weather as the water vapor in the exhaust freezes, making an opaque white cloud. When a truck gets underway, it comes out the side and blows behind. On a plane, the propeller blows it aft.

My pilot said, "Warm up the plane."

I replied with a confident, "Yes, sir." I made a preflight check, got in, put the mixture in full rich, carburetor heat on, and gave it seven shots of prime. Turning the ignition to 'both' I hit the starter. The propeller blade I was watching moved about a quarter inch. The oil in the engine had congealed and the battery was feeble from the cold.

We had a Herman Nelson heater there, but I hated to use it. They burned gasoline, were hard to start, and took time to set up. I thought about the engine's two ignition magnetos. They were made by Bendix, had platinum breaker points and mica insulation. If anything would make sparks, they would. I asked Jim Wright to sit in the cockpit to work the controls and went around to the propeller. I said, "Switch off, throttle cracked, and no prime." I gave the propeller a mighty pull and it made a viscous slow third of a turn. It felt as if the whole crankcase was full of butter. I pulled it through several revolutions, got the propeller in position for a good pull and called out, "seven shots of prime, throttle cracked, and switch on." Jim repeated the words and I gave it a good pull. The two exhaust pipes began to talk to each other- 'puckety puck, puckety puck' and so on. By now several people had gathered and they all cheered. Some good airplane. It took a half hour to get warm enough to fly.

That winter at Big Delta, with the days being so short and cold, there was not a lot of flying going on. We did have a nice ski hill nearby that was originally set up for officers to learn to ski and for Arctic Test Branch to get experience with cold weather clothing, footgear and skiing equipment. A rope tow had been set up to pull skiers to the top and the hill had a broad face with several trails. The rope tow had been powered by a LeRoi four cylinder industrial engine of about forty horsepower. It had seen better days, having frozen oil added by stuffing broken chunks of it into the crankcase too many times as it takes time to liquify and circulate.

We had a good motor sergeant who could requisition things with facility. One day a cargo plane flew in with a brand new Jeep engine in a crate. Prior to this, ATB had been assigned to test a large cargo parachute capable of lowering a Jeep to the ground right side up and ready to go. They tried it with an old Jeep so as to not risk anything of great value. They took it up several thousand feet in a C-119 and pushed it out the cargo door. The chute opened, but it 'streamed' (did not billow out) and so had no retarding effect on the Jeep. Everything in the Jeep was bent or broken except the gear transfer case. This unit goes behind the transmission and gives a high and low range of speed plus having a fore and aft output shaft normally going to the front and rear axles. We were able to use this unit to adapt the new Jeep engine to our rope tow.

An engine driving a ski tow must have a speed governor, so as to maintain a constant rope speed regardless of how many people are hanging on. The old LeRoi engine had one, but the Jeep engine didn't. Other people had been assigned to install the new engine on the rope tow and I had been borrowed from the air section to adapt the new engine.

I fashioned a speed governor for it by taking a door hinge, enlarging one of the holes to fit one of the cylinder head bolts near the front and just behind the fan. Then I made an air vane about four inches square and screwed to the moveable upper part of the hinge. I made an adjustable link from the air vane to the throttle lever out of welding rod and used a screen door spring to hold the throttle open. This way the spring urged the vane up vertically behind the fan. The fan, being belted to the engine, blew the vane back, closing the throttle if the fan went too fast. If the fan was turning too slowly, the spring would pull the vane back up, opening the throttle. The engine quickly reached governed speed.

The people working on the ski tow made a new mounting for the engine. Then they adapted the rope sheave and after all the arguing about details, made it work pretty well. All this was going on in the early spring of 1953.

I was to get out of the Army in May and, since I had a practice of keeping busy, went about making battery chargers. Batteries, whether they be for automobiles, trucks, or aircraft, all suffer from the cold. All of these batteries are of the lead-acid type and low temperatures reduce their chemical activity. When this weakness is coupled to the fact that the engines the battery is trying to start are more difficult to turn over due to the oil being viscous, and the ignition sparks feeble, there are always batteries in need of charging. My best charger used a 24 volt truck generator driven by two Briggs and Stratton engines coupled together by a flexible joint. The joint coupled the output shaft of the first to the front of the crankshaft of the second. They fired alternately and together had enough power to bring the generator to full output of fifty amps.

My urge to keep busy stemmed from my constant desire to become a civilian. I believed that military life alone would not prepare me for that. I simply kept looking forward to May.

In the spring of 1953 I was honorably discharged from the Army and began to assess my prospects for becoming a civilian. My only real enjoyment of, and success in college had been the study of aircraft and helicopters with some stress analysis and aerodynamics. I added that to my experience with them in the Army and made a rather quick assumption that I should find a job in that area.

However, in the back of my mind was a surreal notion of becoming an engineer that never quite left. I had met engineers and worked with them in some early part time jobs and respected their understanding of the natural laws. Those were uniform and utterly reliable. However, my earlier failure in complex math courses made me hesitant to try further college.

Hiller Helicopters had a manufacturing plant in Palo Alto and in early summer of 1953, I got a job there working on their assembly line. We were also doing final adjustments of the rigging and making run up tests. While I was there, I was told I would have to join a machinists union or leave the job after one month. After work that day, I went to the union hall and met the leader of that branch. I asked him what would be the advantage of belonging to the union? He replied in a gravelly voice, "If ya ever out of a job, I getcha a new job." I thanked him for his time and walked away while deciding I would seek any and all future jobs on my own merit.

At the time I got the job at Hiller, I rented a very nice studio apartment in Palo Alto and preferred to live there for some time. Therefore, when my month at Hiller was up, I found another job at Palo Alto municipal airport. The job there consisted of refurbishing a PBY aircraft for use by Campbell Soups. The story was that they wanted it to fly to and from the Hawaiian Islands to tend vegetables and beef for making soups.

The PBY was a large amphibian flying boat that first flew in 1935. The high broad wing had a span of 104 feet. Production stopped later in the thirties after some 3300 were built. They were widely used by the Allies in World War II for submarine spotting, air-sea rescue, etc. The two engines, mounted close together in the center of the high wing were Pratt & Whitney R-1830's of 1200 horsepower each. They were the same as used on the redoubtable Douglas DC3.

Part of my job was to prepare the engines for running. That included taking out all the front spark plugs, spraying into each cylinder a mixture of kerosene and lube oil and pulling the engine through by hand many times. This would free any stuck rings and prelube the pistons. It also left a sizable amount of oil in the bottom of the exhaust manifolds. Then remove the rear spark plugs and clean and set the gaps on all fifty six of them. Others would attend to other needs- replace a de-icer pump, drain and clean the oil sumps, check hydraulic systems for leaks, etc. On this aircraft, the exhaust stacks exited through the cowls above the wing and pointed a little aft and toward center. When starting a radial engine that has been dormant for some time, there should be a fire guard with a large heavy carbon dioxide fire extinguisher behind and near the exhaust outlet.

Since I was young and agile, that task would be mine. I fashioned a rope harness and managed to tie myself and the heavy CO_2 extinguisher about in the center of the wing. This put me between and just aft of the two five inch diameter stacks. When all was ready, the man in the cockpit yelled "Clear one." I yelled back "Clear one." There was a faint whine of the primer pump, then the big whine of the starter. I saw the propeller turning and then "Blouff." About a gallon of oil had collected in the bottom of the exhaust manifold and now it was all over me. Since the left engine was running, nobody could yell and be heard. Therefore, I simply checked my footing and watched the right propeller. Sure enough, it began turning and again "Blouff."

In winding up the restoration of the PBY, the new owners wanted it painted bright yellow. After about ten days of preparation and spraying, we had created the biggest bright yellow thing I had ever seen.

As I was driving to my apartment that evening, I began to think, "An engineer would never have to do that," and if I went back to college, I could get $110 a month on the GI bill. Fall was approaching and I began to get serious about going back to college. I still nurtured the seed of an idea that I might be an engineer. My experience with basic training in the Army showed that, with discipline, I could physically do far more than ever before. Perhaps enough extra effort might be applied in the mental department that I could get a bachelor's degree in engineering.

I had known some students in Davis High School whose parents were professors at the university there and became convinced that U.C. Berkeley would be the best place to study engineering. Also my father had done so. To that end, I took an extensive aptitude test there that cost $45 and took two days. The gentleman that gave me the test was a Dr. Wiseman and was very sympathetic to my intention. The two day aptitude test showed me to be exceptionally good at visually solved problems, spatial relationships and those puzzles wherein if you turn shaft A clockwise, lever Z will move up, down, or reciprocate sinusoidally. Other problems involving higher mathematics or rote memory left me totally in the dust. Dr. Wiseman suggested (and quite rightfully so) that I attend an engineering school less theoretical and more practical than U.C. Berkeley.

My high school transcript foretold no particular ability and my experience at Sacramento Junior College indicated nothing more than that I was a good aircraft mechanic. Therefore, I thought, a bachelor's degree in mechanical engineering would be nothing short of astounding.

Unfortunately I believed my newfound will and determination to be up to the task and continued to pursue the Berkeley route. Fall was approaching and I talked to the director of admissions. He would not let me enter Berkeley directly but suggested I do my lower division engineering at City College of San Francisco. Both he and others told me that it had a better curriculum for lower division engineering than U.C. Berkeley anyway- more practical and less theoretical.

So, in the fall of 1953, I applied to CCSF and was accepted. I found a boarding house near there with three other students. Two of them were also doing their lower division engineering requirements for U.C. Berkeley. The other was in pre-dental. I made up some of my high school deficiencies there and took high school algebra two more times, getting a D and a C. That summer I got a room in Berkeley and took high school algebra for the sixth and last time, getting a C again. During the second summer session there, I took high school trigonometry. I got straight A's as long as it dealt with triangles and picture problems. However, the last third of the course was only about identities. This concerned only numbers and terms and involved no pictures resulting in a C for the course. I put in another year at CCSF completing more lower division engineering requirements with moderate grades. I began to doubt my ability to become an engineer.

In 1955, I began the first summer session at U.C. Berkeley taking physics 4A, a course that made some students tremble. I had an excellent instructor named Earl Frisen. He dealt with real physical experiments in mechanics, heat, and sound, using only enough math to predict the results. Thus we could see how the math connects with reality. I was very happy to use the math for real situations and grasped the relationships between it and visual events. I pulled a B in the course.

Somewhat heartened, I signed up for Organic Chemistry the second summer session, another requisite for lower division engineering. I learned some amount about molecular structure and particularly of the complexity and variety of organic molecules. At one point I understood what a stereoisomeric molecule was. On the final exam was the requirement to memorize the formula for 104 complex organic compounds and write them down. I simply couldn't do it. At that moment, I decided NO MORE COLLEGE.

Since this decision not to try more college was final, I again went over my past to sharpen my aim on a future. I recalled my high school days of working in my father's meager shop. In it I had repaired and made parts for dozens of household items as well as my first working steam engine and first running gasoline engine. They were made entirely from junk and to see them come to life instilled in me an iron-clad conviction that I could design things that worked.

There was a part of school that I made good in- it was drawing, drafting, and mechanical design. In those courses I always did well- from high school, summer jobs, through junior college as well as in those two years at City College of San Francisco. I was confident that I could do well at what I then called an engineering apprenticeship. Surely a person with all these prior experiences must be of value somewhere. Then, what sort of organization might pay me for my efforts? I guessed that all mainline manufacturers would only consider college graduates. This meant that my best bet was with small manufacturers of individual and varied jobs rather than mass production. I also wanted to stay in California where such jobs might best be found.

The next day I put on a suit and tie and went to San Francisco to its Chamber of Commerce. I told the people there of my intent and was shown a huge ledger in which all established manufacturers in California could be found. I came back to Berkeley that evening with a list of twenty nine possibilities.

Since I had a nice place to live in Berkeley, I decided to start looking in that city. Within Berkeley, I amassed five offers to consider. I talked with their chief engineers individually, and settled on a company that made food processing equipment. There the jobs would be varied and would entail a broad spectrum of engineering requirements- good training. I got along well with the office staff and, through summer jobs, had become familiar with most of the processes in the fabrication shop. I made it a point to succeed since I was determined never to go to school again. Also experience had shown me that on-the-job-training could really stay with me since it could all go into my mental Rolodex.

In the process of making cooking and distillery equipment, a great deal of copper work was involved. Copper was favored for large cooking vessels because of its great heat conductivity. This provides uniform surface temperature and freedom from scorching. It can be formed into many intricate shapes because, although it work hardens, it can be repeatedly annealed by heating to a dull red. Steam jacketed copper kettles are excellent for cooking jams, jellies, soups, sauces, tamale filling, etc. Copper's use in distilling equipment is legendary. Since that time (1956), stainless steel has replaced copper in some applications since it is easier to keep clean.

My first assignment there was the design and drawings for a copper brew kettle and its associated equipment. It was to be installed in a brewery in Azusa, California. The brewery there already had three in operation, and since my company had no up to date drawings of the kettle itself I was to go there and make measurements of their existing units and the building's dimensions to accommodate the new one. This brew kettle was the traditional apple shape and was three stories high with a capacity of 320 barrels, each barrel being 22 gallons.

I loaded my car with many measuring devices, a camera and names of the brewmaster and his associates. I had a friend and his wife who lived in Granada Hills west of Azusa. He was an engineer for Lockheed and she was a grade school teacher. I had been best man at their wedding prior to that time, and we agreed that I would stay with them for about a week. As I drove down there I developed a glorious feeling- now I was functioning in the real world and would be answering to people who were really making things happen- learning to be part of something truly significant, something that worked.

The company I worked for was located in West Berkeley in an industrial strip that included Emeryville and part of Oakland. In this strip was a brass foundry, iron and steel foundries, a large pattern works and forge shops. Nearby in Oakland was a machine works that had a lathe that could machine shafts fifty feet long. It had turned out propeller shafts for ships, some of which were built in nearby Richmond. All of the large fabrication shops and metals suppliers had railway spur lines connecting to north-south mainlines. The local fabricators were thus able to ship in large parts for what they were making. Big forgings, castings, flanged and dished tank heads, pieces for oil refineries, heat exchangers, etc.

One of the engineers where I worked designed all of the heat exchangers that we were to fabricate. He showed me many methods, in their design, that would improve performance as well as keeping costs down. As I continued my work there, I became convinced that it was my engineering apprenticeship and (in my particular case) that no college could match it for value.

I truly enjoyed improving my engineering education with practical experience. I discovered that my slide rule alone was sufficient to solve all the real mathematical problems I encountered.

I had always liked music and, although my reading problems prevented me from reading it, I had a good ear. I played B flat trombone with the school orchestra from the sixth grade through high school. The trombone is a middle range brasswind horn that has a slide and, in some cases, valves to vary the length of the air column controlling the pitch of the note played. Therefore the player must have a good sense of pitch, and tone deaf players may be advised not to take it up.

Since my ability to design new devices was improving and I liked music, I figured out a self challenge. What if I was to build a miniature pipe organ of only three octaves that would stand on a table?

Somewhere when I was a boy, I heard a steam calliope playing ragtime music and was delighted by its loud determination to nearly achieve happy music. Steam calliopes were always a little out of tune because if a pipe played a short note, the steam blowing it would be cold and dense, making the note slightly flat. If a pipe played a long note, the steam would get hotter, making the note slightly sharp. This gave a comic effect as though the device could not quite make it as a true pipe organ but was trying hard. At least it was loud.

The steam calliope was invented and patented by J. C. Stoddard of Massachusetts in 1855. He liked the sound of locomotive whistles. Calliope was the Greek goddess of music, song, and dance, and was specifically named as the Muse of Epic

Poetry. Steam calliopes became the signature sound of riverboats and circus parades.

After work in the evenings in my Berkeley apartment I designed this little calliope-like organ. It had a single rank of brass tubular pipes starting at middle C and going up three octaves. The pipes would be a stopped type, tuned with a moveable cork in each one. They were made from brass tube and mounted on top of a four inch square wind chest with a magnetically actuated valve under each pipe. The wind chest was mounted atop a mahogany case that contained an early Westinghouse quarter horsepower washing machine motor. The motor was belted to a rather noisy air blower and a small twelve volt automotive generator. I bought part of a piano keyboard from a widow whose late husband had been a piano tuner. I arranged each key to press a bronze leaf spring against a grounding bar to energize the valve of the corresponding pipe.

Rollin next to the calliope-like organ that he designed and built.

I did most of the fabrication work in my father's small shop in Davis, California, working on weekends. My younger sister Elinor lived there and was a senior in high school. She had become very adept at the piano and loved to play ragtime tunes like those popular in the early nineteen hundreds. After I spent some time tuning the organ, we put it on a table of convenient height. Elinor began fooling around on it and was delighted. She quickly became versatile with it even though its range was only three octaves. It was as though she and the machine had a mutual talent. She could play it in a way that was loud and raucous and could elicit mirth from the most solemn of people. The pipe organ eventually found its home in a pizza place in San Francisco.

In my weekday job as well as in building the organ, I realized that I could design and build a great many things without need of dreaded higher mathematics. These experiences opened a great door that I had previously believed was closed.

When I had been designing food processing equipment for four years in Berkeley, the company got a contract from the Lawrence Berkeley Radiation Laboratory for a large vacuum chamber with windows and an end door. It was my first experience designing a vessel for high vacuum. It would require radiographed welds and a helium leak detector test. At home I studied with the aid of new books, methods of achieving and maintaining high vacuums, and came to realize it is sort of an art. Doing the drafting and engineering on this vessel for the Radiation Laboratory stirred my interest in that facility to such an extent that I applied for work there.

I talked with the employment manager and gave him a written description of my prior experiences and named people for references. He sent me to see the manager of the lab's mechanical engineering department. That person viewed me favorably and started me in the position of design draftsman. There were two classifications up from that so I had room to move up even without a degree. I soon began to meet and work with graduate engineers and even post-doc physicists. We got along beautifully for the most part and they didn't care a trifle about my background. They were mindful of what we could accomplish together, how well it would work and could we keep it under budget. I sought the help of older graduate engineers who had a lot of experience and learned a great many practical solutions to unusual problems- for example designing joints for high vacuum welds, edge effects of lubrication in journal bearings, when drafting a right and left member for an assembly make them identical so the shop can't make two rights or two lefts, if making a safety backstop for a heavy moving part- make it a number of easily replaceable frangible pieces, etc.

For stress analyzing all manner of beams, shafts, and tubes, one fellow gave me an old college text of beam diagrams for finding fiber stress in any beam application. Another wrote down on a single piece of paper formulae for finding the section modulus of any cross section- I still have the paper. In solving problems that went beyond these basics- thermodynamics, heat transfer, analogies between thermal, electrical and hydraulic systems, Ohm's law, etc.- our staff had experts in each and all of them. Very few were stuffy or holier than thou. In fact, the real experts seemed more than willing to share their knowledge and enjoyed the comradery.

Here I noticed characteristics of people who are exceptionally knowledgeable: they tend to talk quietly, have no interest in showing others how smart they are and are endlessly curious. One such person I met shortly after I joined the Lab, who was twenty years my senior. White haired and soft spoken, he and his wife were endlessly rebuilding a lovely old home on Panoramic Way- a narrow meandering road high on the west edge of the Berkeley hills overlooking Berkeley, Oakland and San Francisco.

I mentioned that I had been an aircraft mechanic and he told me of some of his earlier flying experiences. He had owned a Buhl Pup- an early midwing light plane with a three cylinder Szekeley radial engine. One day he blew a cylinder off (which was all too common with them) and repaired it the recommended way by putting three pair of turnbuckles between the cylinder heads in addition to the existing cylinder base studs. On another occasion with the same craft, the propeller came off and fell in a field. Trimming the plane for the loss of weight in the nose, he landed in the same field. After hitchhiking home, he returned with a new shaft nut and key, put the prop on and flew it home. I expressed a shadow of doubt at which time he showed me photographs of those and other calamities. He had, in addition to a world of experience, an engineering degree from a midwestern college. While there, and needing money, he rode motorcycles in one of those huge open top barrels at a carnival. He was highly regarded by others in the engineering department and a mentor to me as well as others who practiced the art of listening.

There were others there that I learned a lot from. One showed me the analogies between heat flow, fluid flow and electrical flow through various systems and what the various members of each system were represented by. I learned how stress-strain curves for various materials could be used to avoid fatigue failure. On a great many occasions I was able to utilize mechanisms and devices that I had stored on the Rolodex. The Geneva escapement movement in watches, the three types of Watt linkages (James Watt) that produce linear motion using only pivoted links, a mechanical differential using only spur gears (no bevel gears) to fit into a very narrow space, intermittent motions for remotely locating targets on a collimated radiation beam- to name a few.

There were a great many problems I solved in my thirty three years there in which the solution was suggested by experience with a similar thing in my past. I cannot imagine coming up with solutions found in college because I didn't get that far into it.

When I had been at my job at the Berkeley Radiation Lab for about four years, one of the engineers showed up for work at the lab on an electric bicycle. He had taken an old twenty six inch balloon tired bicycle and hung a twelve volt automotive battery on each side of the rear wheel. Then, after making a ring sprocket fastened to the rear wheel rim, he chained it to a twenty four volt series wound DC motor. The motor came from a military surplus store which was fairly common at the time. Using twelve volt starter relays, the batteries could be connected in parallel or series for two ranges of speed.

Starting at the northeast corner of the U.C. Berkeley campus, Hearst Avenue was steep and winding for about a mile, ending at the laboratory complex where I worked. When I first saw the quick silent manner in which Pete's bike came up that hill, I began to plot how I might build a light motorcycle powered by two large automotive batteries.

In 1948, Harley Davidson had put on the market a lightweight 125 cc motorcycle. It was a lot heftier than a standard bicycle, having bigger tires and more effective brakes. Another friend at the lab had an old one of these lightweights that was beyond redemption. I was a little unsure about starting a project with no guarantee of good results, but curiosity won and I bought this old bike for thirty dollars. I sold the engine and transmission to another friend for $15.00.

Down the street lived an older man named George Hall. He and his wife had built and were renting units in a lovely apartment complex that had a good machine shop. Previously I had helped him restore a huge 1930 Rolls Royce. He had bought it in 1941 for only three hundred dollars. Why so little? In 1941, the war had begun, gasoline would be rationed and the Rolls' eight liter engine was simply too thirsty. In return for my help with the Rolls, George would let me use his shop to build the improved electric motorcycle.

I laid out the motorcycle on paper locating the battery pair low and forward to obtain a good center of gravity. The frame was to be a box type welded up from steel tube. Just behind the batteries was a large space to mount the motor so that its drive sprocket would line up with a large rear wheel sprocket. Just behind the motor space was the anchor point for the rear wheel swingarm. I obtained a motor from a surplus catalog that turned out to be somewhat feeble and could not provide regenerative braking. I began to study direct current motors at the U.C. library and became acquainted with an electrical engineering professor who worked at the lab.

After some study, I developed a motor that would deliver very high starting torque and variable speed control without the use of resistors. It also provided variable regenerative braking. I experimented with this type of motor by rewinding heavy duty automotive four pole generators. I obtained magnet wire and other supplies from local motor rewinders.

The lab where I worked had a good purchasing agent who told me where I could buy almost anything I might need. My motor speed control system worked so well that I considered applying for a patent on it. Then I happened to tour a World War II submarine named the USS Pompanito which was tied up at San Francisco. It had the same motor speed control system I had developed with the addition of reverse.

The electric motorcycle became well enough known that I began to give talks on it. I was married by then and my wife and I had bought a house in Kensington just north of Berkeley. The house included a separate two car garage in which I set up a shop of my own. After my wife Millie and I had settled in, I

Rollin's homemade electric motorcycle.

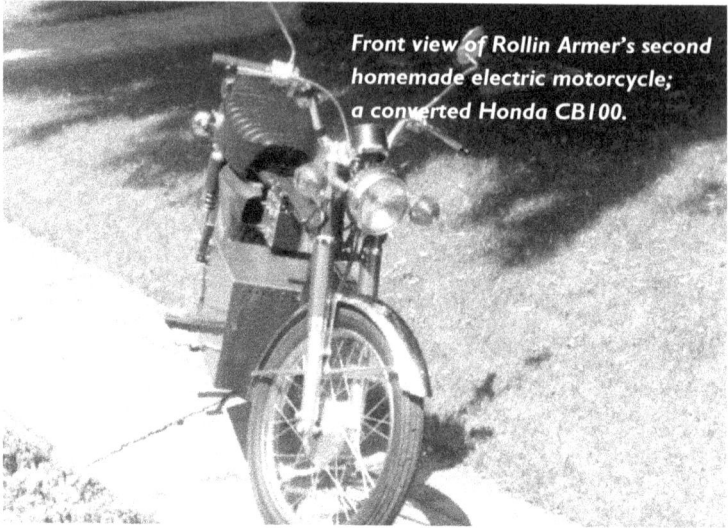

Front view of Rollin Armer's second homemade electric motorcycle; a converted Honda CB100.

began to consider building an electric car.

My early training as an aircraft mechanic along with my four years of riding the electric motorcycle to and from work gave me some insight as to how such a car might be built. I recalled in detail how one all-wood airplane had been designed and built. It debuted in 1929, was a good performer and earned a reputation for safety and arriving on time. It was the Lockheed Vega, a high wing monoplane. It cruised at 130 miles an hour carrying six passengers with a 300 horsepower Pratt & Whitney engine (8).

Also my friend George Hall had built a twenty six foot sailboat all of wood which we used to sail on San Francisco Bay. He favored plywood construction for the car and suggested it should be marine grade. That used waterproof glue and could be obtained with a mahogany face. This way the whole inside of the car was varnished mahogany. Many said it looked like a boat.

I chose to use the chassis parts of some small car. After some study in a foreign car junkyard, I settled on a 1960 Morris Minor for my chassis parts. It had torsion bar front suspension, rack and pinion steering, and an appropriate rear axle and tire size. The combination would be light and strong- no heavier than necessary but strong enough to take the high accelerating torque of the electric motor. Also the six hundred pounds of batteries could be carried on the floor giving a low center of mass and very good cornering ability.

The most successful motor for the motorcycle had been a rewound heavy duty automotive generator. Therefore, for the car motor, I located a very large bus generator that weighed 140 pounds. This one was made by Delco-Remy for use on buses. They needed a big generator because they spent a lot of time idling with all the lights on.

The head of the Electrical Engineering Department at U.C. Berkeley, the late Dr. Arthur Hopkin, had become interested in my electric vehicle activities. He gave me the use of the dynamometer laboratory at Cory Hall to run tests on my car motor. He also assigned a resourceful graduate student named Francis Sakomoto to assist in the tests.

We worked in the lab in the evenings, and in two weeks, had measured the characteristics of my redesigned car motor. It developed twenty four horsepower with an efficiency of eighty seven percent.

After a seemingly endless amount of work on the car, much of it possible due to my patient wife, I took it for a drive. I was delighted. It had brisk acceleration and would climb any hill in Berkeley including Marin Avenue. The cornering and general handling were very good due to the low center of gravity. However, due to using the primitive lead-acid type battery, the range was limited. The farthest I drove it on a single charge was 37 miles in light traffic. Top speed was 50 miles per hour on level road.

Since this car used the old lead-acid battery like the electric cars of yesteryear, its performance was only better due to lighter weight and a more efficient motor and control system.

Since 1970 when that car began its useful life, enormous improvements have been made in batteries, motors, control systems and electric power sources. An electric car today can go twice as fast, ten times farther on a charge and get all its power from the sun or wind.

Today we see huge established organizations plundering our earth for oil, starting conflicts for more and exacerbating global warming. If a long and healthy future for man requires the collapse of one major facet of our economy, wouldn't that be an ultimately sensible trade?

Rollin with his handmade electric car, decades before modern electric cars like Tesla.

Rollin with his homemade electric motocycle and his electric car made of wood.

Why I believe in science and the steadfast natural laws

The laws involve chemical reactions, the behavior of heat and light, mechanics, electricity and magnetism, fluid flow, and the structure and behavior of molecules. The laws have been in place since the beginning of time and will remain into the future far longer than anything man can foresee. They are not affected by opinion or wishful thinking. People who understand them can converse and plan with others of similar understanding.

If, for example, a high speed train is planned that must negotiate a turn of a given radius at a speed which the natural laws predict is too great, the train will leave the track. Further, the kinetic (stored) energy (mass x velocity^2) will be converted to heat by bending, scraping, smashing and tearing of material until the heat produced is exactly equivalent to the kinetic energy present at the beginning. This is how an air bag works. The gases in the bag, being suddenly compressed, get hot. The production of this heat absorbs the kinetic energy, or work, of the person's head, giving it a little time to slow down before stopping.

Consider a bullet fired from a gun: The bullet has mass (weight). It is accelerated down the barrel by the pressure of the violently expanding gases of the burning powder charge. The acceleration force is equal to the pressure of the burning gases times the area of the back end of the bullet. This force stops when the bullet leaves the barrel. Then the bullet is carried by its own inertia which is equal to its mass times the square of its speed. It is slowed in its flight by the resistance of the air it is passing through (the work done on the bullet by the air makes it hot). As it hits the target, it will stop when the heat caused by its smashing and tearing of the target is exactly equal to its mass times the square of its speed.

It is reassuring to know that these same laws apply in more

peaceful situations, in fact in all situations involving mass and motion. Some of us have seen a figure skater on the ice coming to a forward stop as she begins a vertical whirl with her arms and hands outstretched. She then has a steady rotational speed. Now she pulls her outstretched arms and hands in and holds them to her sides. As she does this, her rotational speed increases dramatically. This is because a mass in motion tends to remain in motion unless acted on by an external force. The weight of her arms and hands is first turning in a large circle. By pulling in her arms and hands, she forces them to turn in a smaller circle, so to keep up their speed they have to rotate faster. One can bet that as she stops whirling and goes in a straight line, she first moves out her arms and hands to slow down her spinning motion.

This phenomenon is called conservation of momentum. To look at this in a linear (straight line) situation rather than rotary, imagine a car traveling on a straight level highway at sixty miles an hour. The driver shifts into neutral so the engine is disconnected. The car gradually slows down because its momentum is being converted into heat in two ways: The churning of the air through which it is travelling, and the rolling resistance of the tires. Now the brakes are applied. They are a device for converting work into heat by friction. The brakes can turn work into heat much faster than wind drag and rolling resistance. Now the driver presses the brake pedal so hard that the wheels lock. Then the brakes are stationary and stop making heat. But now the tires are scrubbing on the road so violently that shreds of rubber are burning. In one way or another the energy taken to accelerate the car up to sixty miles per hour will be converted to heat before the car stops.

The energy taken to bring the car up to sixty miles per hour came from the part of the heat that the engine converted to work (about 25%) of the burning gasoline. The engine lost the other 75% into the radiator and out via the flaming exhaust gases. In other words it took heat to bring the car up to sixty miles per hour, more heat to hold it there and it will put more heat yet into the atmosphere in stopping. Modern cars produce less heat than very old ones due to better thermal efficiency of the engine and less churning of the air owing to better streamlining.

All the heat just mentioned has less effect on global warming than the heat trapping effect of carbon dioxide in the atmosphere. The atmosphere is composed as follows: Nitrogen 78%, Oxygen 21%, Carbon Dioxide, Argon 0.9%, methane 0.44%, helium 1.3%, neon 4.67%, nitrous oxide 0.08%, and ozone 0.01%. Methane in the atmosphere is far more heat-trapping than carbon dioxide. Estimates vary from 20 times to 80 times. It is also an excellent automobile fuel allowing higher expansion ratios as well as less carbon dioxide in the exhaust.

Carbon dioxide varies in concentration in the Earth's atmosphere over time. In ancient times, over twenty thousand years ago (1), it was 200 parts per million. In 1950 it was 300 parts per million. Now, in 2017 it is 400 parts per million. It is produced by the burning of hydrocarbon fuels in air to form carbon dioxide and water vapor. Green leaves of plants are able to take it, along with water and sunlight, to produce oxygen as well as leaves and wood, which are hydrocarbons. Interestingly, it was this process which, over much time, produced the fuels we have today: oil, coal, methane, peat, wood, and other hydrocarbons.

Historically this burning of hydrocarbon fuel by man for transportation began with Richard Trevithick's steam locomotive in 1804 (2). A very short time ago in terms of the time it took for all that plant material to form the fuel for his engine and all those that came since. An even shorter period of time was taken for the concentration of carbon dioxide in our atmosphere to increase by 33 percent from where it was in 1950.

The preceding evidence makes a statement that man's activities on earth are creating carbon dioxide faster than green leaved plants are converting it back into oxygen and carbon compounds. Not only that, but when man grows corn and molasses-bearing cane to obtain alcohol for motor fuel, that alcohol yields only 60% of the heat per gallon that gasoline does. Therefore if one's car gets 30 miles per gallon on gasoline, it would get only 18 miles per gallon on alcohol.

Today, when we burn fossil fuel in the form of hydrocarbons (oil, gas, wood, coal, etc.) we are using the oxygen from today's air. To gain a notion of how fast we are using it, consider the following. In the U.S. we burn six times as much gasoline per person as in any other country. In 2014, this was 137 billion gallons. When one does the math, that comes out to be 4,344 gallons per second.

How much air (in familiar terms) does it take to burn one gallon of gasoline? Enough air to fill a room thirteen feet square and nine feet high. We in the U.S. alone use up the oxygen in, and discharge the carbon dioxide and hot nitrogen gas that filled 4,344 of those rooms every second of time.

Of the heat produced from this burning, today's car engine converts, on average, some 24% into car-pushing work- the rest escapes into the atmosphere as heat from the car's radiator and from the exhaust pipe in the form of hot nitrogen, carbon dioxide and water vapor. The amount of car-pushing work required is proportional to the car's size and weight. To the automakers' credit, this 24% compares with 14% 50 years ago. Improvements were made by electronic control of fuel injection and ignition timing, multi-valve cylinder heads and by better transmissions, allowing more ratios. A further improvement was the adoption of radial ply tires that roll easier than the old bias-ply type.

However, it must be remembered that all the work the car's engine delivered in propelling the car was also converted into heat through aerodynamic drag and rolling resistance. The car will stop sooner if the brakes are used because they turn some of the car's momentum into heat. It is therefore evident that all of the heat created by the burning of that fuel went into the atmosphere. The fact that some people were carried somewhere makes no difference to the atmosphere.

If we could provide that bit of transportation to those people without so much heat being involved, it would surely reduce the global warming problem. The wind drag and rolling resistance cannot be eliminated but they may be reduced slightly from where they are by improving streamlining and reducing rolling resistance. These two factors cannot be improved a great deal in private automobiles, but they are inherently better in railway trains. Trains can also very much reduce the energy loss in stopping and starting by elevating the station. Here the train must climb coming to a stop and fall when leaving. This phenomenon has been exploited for many decades in roller coasters and is often utilized in present railway stations. The improvement trains make in CO_2 reduction only holds true if the train is fully or almost fully loaded. Also the CO_2 made in getting to and from the train station must be taken into account.

Now to compare the passenger miles per gallon of fuel in a typical automobile on the freeway to the passenger miles per gallon of a Boeing 747: the automobile is carrying four people at 70 miles per hour and burning 2.5 gallons per hour. This equals 280 passenger miles divided by 2.5 equals 112 passenger miles per gallon of fuel. The aircraft is carrying five hundred people at 640 miles per hour and burning 60 gallons per minute or 3600 gallons per hour. This equals 320,000 passenger miles divided by 3600 gallons per hour equals 89 passenger miles per gallon. Jet fuel and gasoline are similar in the weight of CO_2 produced in burning one gallon. It is very close to twenty pounds in either case. If one favors air travel because it makes less global warming than other means, the fuel taken to get to and from the airport and the time lost once there must be factored in.

Here it is seen that air travel causes more atmospheric warming per passenger mile than travelling by car. After adding the time lost in airports, flying on trips that could be made by car loses some of its appeal.

From the foregoing, it can be concluded that electric railway trains (fully loaded) make the least atmospheric pollution per passenger mile. Automobiles are a little worse, especially if they are not full of passengers, and aircraft are a little worse than cars in this regard.

However, electric cars are by far the best, especially if they are charged by solar or wind power.

Man's use of heat to do work

Some may argue that man's first use of heat to perform work was the use of gunpowder to settle arguments and take property from others whose gunnery was less effective. Considered here is the use of heat to produce steam for pumping water out of coal mines in the British Isles (3). Briefly put, the first person to do this was Thomas Savery who, in 1700, made a 'steam' pump as follows: a simple boiler at the top of a brick furnace, connected to two vertical tanks which, by the working of four check valves at the tank bottoms could alternately let some of the cold outlet water fall into it condensing the steam, causing a vacuum. This sucked open the inlet valve and lifted water up the inlet pipe. When that tank was full, a valve at the top connected it to the boiler. The steam pressure then forced the water through an outlet check valve, up the pipe to above ground level where it spilled into a reservoir or a river. A person standing on the operating floor had to actuate alternately the two valves from the boiler to the tank tops. Thus it was the condensing of steam that caused the vacuum which lifted the water.

Two shortcomings of Savery's engine were that the greatest height that water may be drawn up a pipe by vacuum is 33 feet and the boiler, being some distance down in the mine, might be dangerous as a source of ignition for mine gases, as well as taking breathing air for the fire.

A new steam experimenter, Thomas Newcomen (1663-1729), an ironmonger from Devon, was instrumental in developing an improved mine pump. With his helper, John Calley, he devised a huge 'walking beam' engine where the boiler and other parts were above ground. The walking beam was like a large teeter totter whose pivot axle was atop a tower over twenty feet high. One may have seen a single cylinder walking beam steam engine in a ferry boat on San Francisco Bay prior to about 1965. Here was an example of that system.

On the end of the beam over the mine shaft was, connected by a piece of chain, a pump rod going deep into the mine. There a pump cylinder was operated, bringing the water to the surface where it was discharged. The other end of the walking beam was directly over a large open top steam cylinder. Its closed bottom connected to a dome topped boiler by a valve. Facing up into the cylinder from the bottom was a shower head connected by a valve to a water reservoir some distance above. In the cylinder was a piston connected by a rod to the walking beam end. To operate, the fire was made under the boiler and brought the water to a boil but under no great pressure. A valve from the boiler to the cylinder was opened. The pump rod end of the walking beam was heavier so it fell to its lower limit in the well. This brought the steam piston up to the top of its travel filling the cylinder with steam at low pressure. Now the valve from the cylinder bottom to the boiler is closed and the cold shower is turned on. Here the power stroke occurs. The cold shower condenses the steam causing it to shrink greatly, sucking the piston down forcibly and hauling the pump rod on the other end of the beam up, lifting much water. The shower also took heat from the cylinder itself.

Newcomen died in 1733 and by then some number of his engines were operating in Western Europe as well as those in England. Then, born in 1736 was an exemplary Scotsman named James Watt. He became not only an engineer but an avid scientist looking deeply into the behavior of steam. He looked into the mine pumping engines of both Savery and Newcomen and saw, that in both, most of the work was accomplished by the vacuum caused by condensing steam. In Savery's engine, each of two vessels were alternately filled with hot steam driving the water up, then valved to suck up water from below as the steam condensed, forming a vacuum. In Newcomen's engine hot steam was let in to let the heavy pump rod on the far end of the walking beam fall to the bottom of its stroke. Then the shower in the cylinder was turned on to sap the steam of its heat causing it to condense, forming a vacuum which did the work by sucking the piston down and the pump rod up as it was on the other end of the walking beam.

James Watt studied both engines and realized that in each case, hot steam was let into a cold cylinder, giving up some of its heat, then deliberately cooled thus throwing away the rest of its heat and doing work by shrinking to a partial vacuum. Also the operating speed of these engines was limited by the time taken for the heat to leave the steam on each power stroke. Watt's first and most effective invention was the separate condenser for steam engines. Here the working cylinder stays hot so that no inlet steam condenses in it. Rather, the steam expands driving the piston to the end of its stroke doing work. Then, as the piston returns to its starting position, an exhaust valve opens connecting the cylinder to the separate condenser. The condenser is an arrangement of tubes with cold water running through them. Its purpose is to expose the exhaust steam to a lot of cold surface, pulling heat out of it so that it will shrink back to the water it was previously.

The separate condenser cooled and shrunk the exhaust steam outside the cylinder, pulling a vacuum on it, but leaving it hot. This way the steam let into the cylinder could be under positive pressure and very hot as it entered. With this greater heat and pressure, the inlet steam could be shut off partway down the stroke. Then, as the piston completed the stroke, it would convert the steam's heat into work. Steam was conserved since the inlet valve closed early in the stroke, and further work was done on the exhaust stroke as the steam met the cold condenser. This cooling caused it to shrink, sucking the piston back to the starting point. There the exhaust valve closed, the inlet valve opened and the cycle began anew. This also caused the cylinder to stay hot so that no heat would be lost to the incoming steam. In fact Watt also insulated the working cylinder to keep it hot. Now, since the cylinder was to stay hot, Watt chose to work both ends of it by employing a piston rod. This was a rod entering a closed and working end of the cylinder through a packing gland, a small bore through one cylinder head with greased packing around the rod to minimize steam leakage. The rod was fastened to the piston.

This way the piston could both push and pull alternately and produce twice as much work from one cylinder. Now the piston would have partial vacuum on one side and pressure on the other, doubling the work a single piston could do. The double acting feature does not improve steam economy of itself, but does allow the engine to be more compact, causing one cylinder to do the work of two single acting ones. Now, since the piston rod is travelling in a straight line and one end of a walking beam travels in an arc, there must be a linkage of some sort that will connect the straight up and down moving end of the rod to the part circular motion of the end of the walking beam. Not just to connect them, but to transmit large forces from one to the other.

Watt worked out three different mechanisms to accomplish this- each one simple, strong, and wear resistant. Today they are sometimes used in suspension systems of automobiles and are known as Watt's linkages.

In 1775, Watt partnered with Matthew Boulton, a prosperous manufacturer in the Soho district in London. Now, with a working cylinder that could both push and pull, Boulton and

Watt devised an engine with a rotating output shaft. They did this by joining the far end of the walking beam to a crankshaft with a connecting rod similar to those in modern engines. This was done after a more complex system had been devised to avoid patent infringement. Now, with a rotating output shaft, the steam engine could be connected to drive any load previously driven by water wheels, windmills, or horses. The horses had been harnessed to a treadmill or to a rotating sweep travelling in a circle.

Wishing to describe the rate of work an engine could do, Watt coined the term 'horsepower.' He measured this by putting a spring scale on a harness, the other end of the scale connected to the load to be pulled, then measured the horses' average speed during an average shift of work. This way he could state for any engine he sold, how many horses it would replace. One horsepower equals 33,000 foot pounds per minute. This shows how hard it is pulling times how fast it is going.

The firm of Boulton and Watt between 1775 and 1800 produced and installed nearly five hundred engines (3). Arthur Woolf in 1805 developed the two stage expansion engine. James Watt's engines were relegated to low pressure steam partly due to early boiler technology and by his interest in the safety of his products.

Woolf introduced the concept of steam expanding first into a relatively high pressure cylinder that exhausted into a larger cylinder, doing further work before exhausting into a vacuum. It was called double expansion. As boiler fabrication technology progressed, higher steam pressures became feasible. This allowed greater expansion ratios to be utilized and engines to be smaller for a given power output. Richard Trevithick's first steam locomotive in 1804 used forty pounds per square inch pressure. Triple expansion was brought in by Alexander Kirk for a boat in Govan, Scotland.

In the early 1800's, the use of steam power was increasing

worldwide. Most of the improvements at that time were made intuitively with trial and error mixed in, as well as learning from the experience of others. In the early 1800's came a brilliant young French physicist named Sadi Carnot (1796-1832) (3, 4). He founded the study of thermodynamics (the behavior of heat). This study, having a mathematical basis, required a new concept for understanding what goes on in the expansion of a gas (including steam) driving a piston on a power or expansion stroke. The new concept defined the quantity of heat doing the expanding. The new concept was of a temperature where there is no heat. All molecular motion ceases.

This would be the basis of Carnot's mathematical relationships. This zero is -273.16°C or -459.69°F. From this came Carnot's equation for the thermal efficiency of any engine utilizing expanding gases to produce power or work. The equation is $T_1\text{-}T_2\div T_1$ where T_1 is the temperature at the beginning of expansion and T_2 is the temperature when the gases are released (the exhaust valve opens). The working gas may be steam, burning fuel and air (gasoline, Diesel, kerosene) or air as in early torpedoes, mine locomotives, rail cars, rock drills, model airplane engines, and carbon dioxide gas as in mine locomotives (called soda motors), or it may be explosive combustion products as in other torpedoes, nail drivers and large aircraft engine starters.

Since 1840 when engine designers began to benefit from Carnot's understanding of heat, improvements in steam engines became common. Most engines of the time used inlet and exhaust valves in a single bore or valve chest parallel to the cylinder. This way, the inlet (hot) steam entered the cylinder through the same ports, or passages, that the exhaust steam (cooler) left through.

On March 10, 1849, George H. Corliss was granted patent #6162 on his engine where the inlet valves were entirely separate from the exhausts. This way, the inlets stayed hot and took no heat from the incoming steam while the exhausts stayed colder, giving no heat back to the spent steam. Also the Corliss valves were in the form of oscillating shafts with grooves in them. They were actuated by an oscillating wrist plate driven by an eccentric on the crankshaft. By manipulating the linkage to the wrist plate, the inlet cutoff time could be shortened for better steam economy at light load and the engine could be reversed.

Steam locomotives were gaining the benefit of variable valve timing as well. They required great uniform starting torque to accelerate the train from standstill to cruising speed. However, once that speed is attained, it can be held there, not by closing the throttle, but by shortening the inlet valve cutoff time.

On most steam locomotives, there is a large control lever in the cab, sometimes called the Johnson Bar. It regulates a linkage on each side called the valve gear, with gear referring to apparatus, not to a cogwheel. The valve gear controls the timing of the opening and closing of the steam inlet and exhaust valves relative to the piston motion. To start the engine, the fireman has tended the fire so that the steam pressure is normal. The driver opens the cylinder bleeder valves to ensure that no water is in the cylinders as the pistons come to the cylinder heads. Then the Johnson Bar is moved to the full forward position. Then, using whatever whistle and bell signals are appropriate in the area, opens the throttle slowly so that the wheels don't spin and lose their grip. As the engine starts, the bleeder valves alternately hiss and blow clouds of white water vapor. Soon the cylinders are warm and the driver closes the bleeders. The train is accelerating as it leaves the town and there is clear track ahead. The driver knows, that to keep his schedule, he wants to maintain fifty miles per hour. As the train nears that speed, the driver does NOT close the throttle. Instead he eases back on the Johnson bar slowly to where the machine is just holding an even fifty. He lets the bar latch in that position.

Now to look into the engine to see why it is operated in this manner: What if, before the engine was started, we saw that the wheel crank pins were at a position of nine o'clock or three o'clock. Then the piston rod could pull or push forever and not cause the wheel to turn. The reason the wheels will always turn is that if the crank pins are at nine or three on one side, they are at twelve or six o'clock on the other side. This ensures that the power pulses are evenly distributed as the crankshaft rotates. This is true for almost all piston engines, be they steam, gasoline, or Diesel. Two notable exceptions are old John Deere tractors and some motorcycles.

Going back to Carnot's equation for thermal efficiency (to use as little heat as possible), T_1-T_2÷T_1, it may be seen that the highest heat efficiency the engine can achieve is where T_2 is as much lower than T_1 as possible. The locomotive driver leaves the throttle wide open so that, at the beginning of the power (or expansion) stroke T_1 is as high as possible. He then limits the inlet steam cutoff time to as early in the stroke as possible, letting in just enough to maintain the power needed. This way the steam expands all the way to the end of the power stroke when the exhaust valve opens, and causes the difference between T_1 and T_2 to be as large as possible. In the language of the engine drivers, "This is using the steam expansively, not expensively." The fireman likes it too because he must add fuel less often.

Here, we may take a look at the steam turbine. It is the steam engine of the present time and drives most of the electric power plant generators worldwide. The chief difference between the turbine and the piston engine is the rotational speed of the turbine output shaft and the crankshaft speed of the engine. The driving wheels (crankshaft) of a typical steam locomotive built in the early 1900's turn about 250 rotations per minute (RPM) at 50 mph. The rotor of a small steam turbine of equal power might typically run at over 10,000 RPM.

Credit for the invention of the steam turbine goes to Sir Charles Parsons, an Englishman who was granted a knighthood for his contributions to England and the world (5). His first experimental turbine (1884) was to be direct coupled to a direct current electric generator that he designed and built. He believed, from experiment and calculation, that this very small turbine would run efficiently at 18,000 RPM. In order to prevent heavy thrust loads on the turbine rotor, he introduced the steam at the center of two groups of rotor and stator blades where it expanded both right and left, producing no end thrust. So successful was the direct coupled turbo-alternator that in 1923, one was installed in Chicago that put out 50,000 kilowatts (67,000 horsepower).

The British had always been interested in sea travel since they were an island country. Steamships had thus far been driven by multi-stage piston steam engines- and quite successfully since they could use water cooled condensers. This way they could pull some vacuum on the exhaust of piston type steam engines. This would expand the steam further, causing T_2 in Carnot's equation to be much lower. This improved efficiency just as it had in Watt's engine.

While on the subject of condensers improving steam economy and thermal efficiency, it should be noted that this was the failure of the steam automobile. It had only air to cool the condenser. Early Stanley automobiles had no condenser and constantly dribbled water from the exhaust, getting little more than a mile per gallon of water. Later steam cars all had condensers in the form of a radiator in front but were eternally thirsty. What about locomotives? They didn't have a condenser, but instead had a tender car to carry fuel. About two thirds of that tender was make-up water for the boiler.

Before leaving the accomplishments of Sir Charles Parsons, two more should be considered. Steamships of the time had always been driven by piston type multistage steam engines. With their slow rotational crankshaft speed, they could be directly connected to large propellers giving high efficiency. The turbine, however, turned way too fast to drive an efficient propeller.

In order to experiment with turbine driven vessels, Sir Charles Parsons had established a separate works at Wallsendon-on-Tyne. Here, he had built a hundred foot long vessel named the Turbinia. It had three unshrouded propeller shafts with skeg bearings near the stern to keep them running true. After some experimentation, he drove each shaft with one stage of a triple expansions turbine. He fitted each shaft with three small propellers- two forward of the skegs and one aft. Now he had the total power divided up between nine small propellers rather than one large one. This way they could turn at very high rotational speeds without their blade tips exceeding efficient speeds.

There was a great Naval review held in 1897 to celebrate the Diamond Jubilee of Queen Victoria. A huge fleet of Naval vessels was assembled off Spithead. With Sir Charles himself controlling the machinery, the Turbinia whined past the Admiralty at 37 knots. This was 42.5 mph, faster than any vessel afloat.

Another design challenge for Sir Charles lay in the design requirements for the center one of the three propellers on the Titanic. Prior to 1912, the problems involved with gearing for high speed turbines driving huge low rpm propellers for large ships were not yet solved. In 1906, the Lusitania and Mauritania had been powered with 68,000 horsepower geared turbines. Vibration problems with the propellers and gearing were so severe that they had to be rebuilt. Also their coal consumption was excessive and they were refitted to burn oil.

For the Titanic, the two wing shafts were each driven by a four cylinder triple expansion piston engine. Each cylinder had a stroke (piston motion) of seventy five inches. The first stage cylinder had the smallest bore of 54 inches. The second stage, intermediate bore of 84 inches. The third stage of expansion (being huge) was carried out in two cylinders, each of 97 inch bore. The two third stage cylinders were a quarter turn apart on the crankshaft to obtain optimum balance and smooth output torque. The engines were three stories tall, and at full power, turned 76 RPM, each one developing sixteen thousand horsepower. There was an enormous condenser with circulating sea water to create some vacuum on the turbine's exhaust.

Sir Charles' challenge was to design a steam turbine to operate on the exhaust system from the two huge engines on its way to the condenser. The center propeller should operate at moderate rotational speed in order to efficiently convert the turbines' great rotational power into ship pushing thrust. The two wing propellers were 23 feet in diameter and, turning 76 RPM, the blade tips were moving at 5500 feet per minute.

Just as Sir Charles had used tiny propellers on the Turbinia to accomodate high shaft speed, he must, in order to direct drive a very large propeller, make a very low speed turbine. He also knew that, for a turbine to run efficiently, the rotor blades must be running at a very great velocity as they pass between the stator blades. The way to do this is to have very large blade wheels. The turbine rotor was thus made 14 feet in diameter and the stack of wheels was 14 feet long. Running on the exhaust steam from the two piston engines, the turbine made 16,000 horsepower at a mere 165 RPM. The cold wet steam then went directly to the sea water cooled condenser and was stored for boiler makeup. Since steam turbines can't run backward, the center turbine on the Titanic was bypassed for reverse, the exhaust from both engines then going directly to the condenser.

Other early steam turbine developers were DeLaval in Sweden, Rateau in France, and Curtis as well as General Electric and Westinghouse in the U.S. Turbines became common for driving ships when gear reduction systems were developed that could take the high input speeds that were typical. Turbo-electric systems came into use wherein a high speed turbine is directly coupled to an alternator, which is an alternating current generator. The generator then drives an electric motor which is geared to the propeller. This system gives good speed control and can be quickly reversed.

Thomas Edison, Nicola Tesla, George Westinghouse and the argument about Alternating and Direct Current

In the 1870s Edison and his team had learned to make direct current electric generators as well as motors. Collectively, direct current machines were called dynamos. Edison had a boiler and steam engine in his shops to drive the dynamos and so had a ready source of direct current at 110 volts. This was used to power all the early lights that Edison invented and also to power other shop equipment. In 1882, Edison had electrified New York City's Pearl Street Station, illuminated with 1400 lamps. To provide this direct current power were six big steam engine driven dynamos, each weighing 27 tons. The system worked at 110 volts DC.

During this time there was a young Serbian electrical wizard in his own right named Nicola Tesla who was working with Edison. Tesla invented and patented alternating current (AC) generators and motors and the transformer. Tesla saw that the Pearl Street Station, operating on direct current, had to have very heavy wiring to carry the current to all those lights at 110 volts, some over a mile from the source. He believed, and rightly so, that if alternating current had been used, the voltage could be stepped up many times, delivering the power at a much lower current. Thus for power transmission at any considerable distance, if the voltage is high enough, the required power (watts) can be delivered at very low current (amperes). The size of the wire needed is related only to the current it must carry, not the voltage. Volts x amps= watts. Edison believed strongly in direct current for all uses and that alternating current was deadly dangerous. Edison and Tesla argued about this to such a point that, in 1885, Tesla left Edison's organization.

Meanwhile, an entrepreneurial inventor in his own right named George Westinghouse (who had patented the failsafe railway air brake in 1869) saw Tesla's patents and licensed them in 1888. Westinghouse saw immediately the advantages of the alternating current system- that its motors and generators were simple and efficient as well as enabling long distance power transmission using voltage transformers. In addition to the work of Tesla, such transformers had been developed by Lucien Gualard of France, and Werner Von Siemens of Germany had developed an AC generator called an alternator. Westinghouse pulled together the talents of these people and others and in 1889 formed the Westinghouse Electric Corporation. Much of Edison's productivity was taken up by General Electric and it became a competitor of Westinghouse.

Sixty cycle alternating current was decided upon early in the electrification of the U.S. Some of the reasons for that are as follows: In transformers, the iron core is made up of laminations. These are made of iron sheets about 1/32 inch thick. If the AC frequency was higher, these sheets would have to be thinner, thus costlier. If the frequency was lower, they would have to be larger overall. At sixty cycle AC output, an AC generator can run at 3600 RPM, which is a very good speed for a large steam turbine to operate- so that they can be direct coupled, using no gears. At the same AC output a huge AC generator, say something over 50,000 Kw can run at 1800 RPM coupled to a larger turbine, still direct coupled, making sixty cycle power. Thus, it was not an accident that an alternating frequency of sixty times per second was established.

Electric power plants are the plants that supply nearly all of the household and commercial electric power worldwide. They come in four basic types: hydroelectric (falling water), steam turboelectric (including nuclear), wind turbine, and photovoltaic (solar).

Hydroelectric generation is accomplished by any one of three kinds of water turbine. These three types of water turbine today are highly energy efficient. Each has an energy conversion efficiency of over 90%. The remaining 10% turns to heat. Water, unlike steam, is incompressible. Therefore all water turbines convert the momentum of falling water into electric generating shaft work in a single stage. The highest pressure turbine, running on a water source far above the turbine, is the Pelton wheel. This water wheel was invented by Lester Allan Pelton about 1877. Here the water is squeezed down into a round nozzle so that it acquires a very high exit speed. Each 'bucket' on the edge of the wheel looks like two spoons side by side. As the wheel rotates, the spoon pairs sequentially come into the water blast where they are forced away in no uncertain terms. Ideally the buckets are moving half as fast as the water. This way they bring the water's speed down to the bucket speed, converting all of the water's momentum into shaft work. The power is governed and the shaft kept at synchronous speed of, say, 1800 RPM- if the electric load diminishes, the governor closes the nozzle valve- if it increases, the governor opens more nozzles.

The water turbine most adaptable for intermediate heads (or water heights) is the Francis turbine. Invented in 1849 by James B. Francis, it consists of a rotor shaped like two shallow cones, one above the other, spaced apart by about an eighth of their diameter on a vertical shaft. The lower cone has a big hole in the center for the water to leave. In between them are a number of curved blades. Surrounding this rotating wheel is a toroidal (doughnut-shaped) housing surrounding the rotor. Inside the housing are a number of tangential vanes inducing the water to whirl as it moves into the edge of the rotor. The vanes between the cones of the rotor catch the whirling ring of water, trying to go with it and producing great torque as the water moves toward the center and falls out the bottom. It may be thought of as a centrifugal pump operating with the water running from edge to center rather than center to edge, the water driving the rotor.

In power dams where the water height is small, but the volume per time (flow rate) is high, the most effective turbine is the Kaplan, invented by Victor Kaplan in 1913. It uses a vertical shaft with a rotor that looks like a ship's propeller. The propeller runs in a vertical axis guide tube with a reduced diameter waist where the rotor turns. This waist (or venturi) increases the water's downward speed as it engages the rotor, thus imparting the greatest possible torque (rotating effort). Note that all these inventions are quite recent in terms of human history.

Steam Turboelectric Power Plants

These come in four kinds depending on the source of heat. Aside from the source of heat, they are all similar. Since they all work on high pressure steam, all are heat engines. Their efficiency in converting heat to electricity is the highest of any heat engine system. The overall efficiencies of the various systems are about as follows: Coal 32-42%, fuel oil 45-48%, natural gas 50-60%, nuclear energy 33%. Coal fired plants tend to have modest efficiency because they produce ash both in the fire and flying out the stacks. The boiler is unable to extract the heat from this ash. The ash causes big disposal problems. Oil fired plants are more efficient than coal burners because the fuel takes less heat to bring it to combustible temperatures. Also, its greater hydrogen content produces more water vapor relative to the CO_2 in its exhaust. Natural gas is the most efficient, because in it gaseous form, it requires no heat of vaporization to bring it to a combustible state. Also, since its formula is CH_4, it produces much of its heat by hydrogen going to water vapor and less from carbon going to CO_2. A major advantage of burning methane is that it occurs naturally and is a far worse contributor to global warming than carbon dioxide is. Nuclear power plants tend to be less efficient than other types because the reactor itself is a closed system and produces steam by having a metal heat exchanger (heating element) inside a water boiler producing steam to drive turbines. Since nuclear reactors do not require combustion air, they are used on submarines to drive turbines.

On all steam driven turbo-electric power plants, the exhaust steam from the turbines is condensed and becomes boiler makeup water. This way the makeup water is still warm and the heat loss, as well as the water requirement, is minimized.

The condensers are either of two types, water cooled or air cooled. The water cooled type utilizes river or ocean water in a heat exchanger to turn the exhaust steam back to hot water for boiler feed. The air cooled consists of one or more tall vertical structures, open at the top with a series of open shelves inside. The exhaust steam enters the bottom, warming the air causing it to rise in the center and fall all around the edge. As it falls, it makes all the shelves wet thus causing an enormous surface of evaporating (thus cool) water. Some of these are in the form of a vertical concrete cylinder with a waist, while others may look like a rectangular building. There seems to be a myth that the concrete cylinder evaporators signify a nuclear power plant. Not true, they may be used on any steam system.

Generating electricity from the wind

Why don't modern wind-electric rotors look like the old water-pumping windmill? The old pumping type had many blades (sometimes 24). This gave them two qualities necessary for their job. The big wind jamming frontal area gave them high starting torque to lift the heavy deep pump shaft up in the well with a column of water above. They should be able to do this in a light wind. Also, in a very high wind, the numerous broad blades shadowed each other to prevent overspeeding.

The old Dutch four arm type bears mention because it shows the cleverness of some of our fore bearers in cooperating with nature. Each of the four arms had a lattice frame which, by itself let the wind through with very little resistance. At the inner end of each arm by the hub was fastened one end of a long canvas sail wrapped into a spiral bundle and held there by a rope sewn into its center. All four ropes went into the shaft center and fastened to a cleat on the inner end of the shaft. To get the most work out of a light breeze, the ropes were let out. This allowed the sails to roll out to the tip of the lattice blade and be supported by it, putting the most surface before the wind. If the prevailing wind became so strong as to damage either the windmill or the driven load, the sails were reefed by pulling in the four ropes and again making them fast to the cleat. In order to keep the windmill facing into the wind, a boom came out behind the main tower with a little windmill at the end. Looking at it from above, the little windmill axis was at right angles to that of the main one. This way when the big wheel was facing directly into the wind, the steering one had the wind coming edge on and would not turn. When the wind shifted, the little wheel would turn the appropriate direction to face the main wheel back into it. Man's use of windmills probably began when sails began to propel boats.

Today, as we seek new sources of power less destructive to our environment than in the recent past, the windmill is making a comeback. Modern wind turbines have evolved into a standard type- the wheel size and tower height vary to make the most of the prevailing wind conditions. The energy in the wind varies as the cube of its speed. Thus a wind wheel will produce eight times as much power in a forty mile per hour wind as it will in a twenty mile per hour wind. This behavior makes the finding of good locations for a wind farm very critical. Also the minimum wind speed to produce any useful output is 13 miles per hour.

Now for a look into the economics of the steam driven electric power plant- both monetary and heat producing. Wind turbines have some of the same economics, but their fuel (the wind) is created by temperature differences in the atmosphere. The power company produces the electricity and sells it for a profit. They buy the equipment as well as the fuel to make it operate. Therefore they look to buy the equipment that will give the greatest output in terms of first cost, fuel cost, transmission cost and lowest downtime and maintenance cost. The makers of all the components that contribute to the plant and its operation, ideally, compete with each other to produce equipment that will most closely fit the power company's needs. Unfortunately, the overall heating of the earth's atmosphere by steam driven electric power plants is very similar to that of the automobile. The automobile burns fuel with air to make it go. All of its engine's heat losses go into the air as heat from the radiator and from the exhaust. All of the engine's work output goes into heating ambient air by churning it, heating it from the brakes and flexing of the tires.

All of the electrical output of a power plant eventually shows up as heat in the atmosphere in the following ways: the total power grid with its miles of overhead wires and transformers, electric motors and all that they drive- air conditioners, fans, all manner of machines, electric irons, stoves, water heaters, dryers - anything that uses electricity eventually turns it into heat.

Electric and hybrid-electric vehicles stand alone in their ability to recover some of the electric energy taken to propel them by regenerative braking and use it later. It is encouraging to see the enormous improvements that have recently been made in many electrical devices.

Electric lighting takes but a tiny fraction of the power it once did. Storage batteries give over twenty times the power per pound both in terms of output rate and quantity per pound than they did forty years ago. Due to new electronic switching devices and magnetic materials, today's electric motors are far smaller, lighter, and more efficient than we would have dreamed of forty years ago. A drone aircraft, flying by remote control, carrying a payload many miles and returning on a charged battery seems, to this author, an astounding example of people cooperatively using their heads. All of these factors combine to make the electric car, as well as the hybrid, a much more reasonable proposition than they were forty years ago.

New fuels like natural gas and other hydrocarbon gases that exist in our atmosphere at this time are as follows: methane (CH_4), ethane (C_2H_6), propane (C_3H_8), butane (C_4H_{10}), and pentane (C_5H_{12}). Any and each of these can be and has been used as automotive fuel. They all exist in the earth's atmosphere to a more or less tiny amount with the exception of methane which is constantly being produced.

Methane can be produced by microbial breakdown of plant fiber. This takes place often in swamps, garbage dumps, and in the intestines of animals that have a diet rich in this fiber. Some African termites that build nests over six feet high above ground have members dedicated to fanning lower entrances with their wings to drive in air to displace the methane and drive it out the top (1).

Cattle often go out in open fields to eat grass. When full, they usually retire to a shady place, lie down and chew their cud. They have four stomachs in series to slowly break down the grass that has been rechewed over and over to a fine pulp. This enhances the digestive process which also produces some methane. Methane, hereby is collected from cattle feed lots. By covering garbage dumps with an airtight sheet, methane is collected and used as motor fuel. Rice growing, too, is a significant source of methane. Also methane bubbles up in the oceans of the world. Methane is held at great ocean depths as a gas hydrate- a solid form of water containing gas molecules in its molecular cavities. The gas is usually methane. It is stable at depths of 1600 feet and over. This hydrate is a crystalline structure of water forming a solid similar to ice called methane clathrate. A clathrate is a substance in which a molecule of one compound fills a cavity within the crystal lattice of another compound. If the crystal lattice is ice, the captured molecule (in this case methane) will be released when the ice melts.

Warming oceans now make more giant plumes of methane in the Pacific northwest. About one quarter of global warming being experienced today is caused by methane emissions. The television show Planet Earth said in 2012 that if the Antarctic ice sheet collapses, billions of tons of methane could be released causing dire global warming. At the stable depths of 1600 feet, the seawater pressure is 703 pounds per square inch. If a gas bubble an inch in diameter is released at that depth, it will grow to over 3.5 inches in diameter as it reaches the surface. With this in mind, one can imagine the volume of gas at the surface that many plumes would create.

Research from British Petroleum and Texas A&M University estimate underwater reserves to be from 35 to 177 quadrillion cubic feet. Proven world reserves from other sources are only 6 quadrillion cubic feet. These facts should provide a large incentive to prevent temperature rise in the world's oceans.

Morgan Downey in his excellent and informative treatise on oil and gas in and on the earth 'Oil 101' explains methane hydrate this way: "Methane hydrates are molecules of methane surrounded by, but not bonded to, molecules of water. Methane hydrates are crystalline in appearance and have traditionally been perceived as a nuisance as small amounts occasionally clog natural gas pipelines. The estimated naturally occurring quantities of methane available from methane hydrates are enormous, more than all coal, oil and non-hydrate natural gas combined. Naturally occurring methane hydrates are most often found under ocean floors. Producing large quantities of methane from methane hydrates in a commercially viable manner is not currently technically possible. One danger with methane hydrates is that an accidental release of large quantities of methane into the atmosphere could rapidly exacerbate global warming. Methane is up to ten times more effective at trapping the sun's heat than carbon dioxide." This estimate of ten times was made prior to 2009 when the book 'Oil 101' was published. More recent estimates vary from 20 times to 80 times more effective at trapping the sun's heat.

One reassuring fact is that methane, being lighter than air, rises into the upper atmosphere. There it is eventually oxidized, producing carbon dioxide and water. As a result, methane in the atmosphere has a half life of seven years. Here is seen a self-sustaining system wherein methane is released into the atmosphere at a certain rate. This rate increases as the earth's atmosphere warms. The rising methane is converted into carbon dioxide and water also at a certain rate. If the rate of release of methane exceeds its rate of oxidation in the upper atmosphere, then global warming will increase. If that happens, the system will feed on itself and the earth as we know it will, in a short time, become unlivable.

This scenario is based on facts collected by numerous groups of individuals, each one of them having spent a good part of their lives in an effort to understand what nature has to show them. Sometimes one does not want to believe them, but eventually does, if what they say makes a great deal of sense or can be demonstrated. Corroboration by several informed sources also leads to acceptance as a fact. It is felt that the best course of action is to extend our understanding of how the systems in play interact and reject any suggestion of systems that don't comply with the laws of nature. In short, take what we know to be true and build on it.

Another aspect of humanity that must have a major impact

on its future is population density. According to an excellent account of recent earth history "The Way Nature Works," modern man is thought to have evolved in Africa thirty five thousand years ago. Spreading across the world over time, he replaced Neanderthal people and survived the end of the Pleistocene ice age. That ice age ended about ten thousand years ago when the world population was about four million people. Six thousand years ago, it was seven million. Then, it really began to grow. Three thousand years ago it became 50 million. At the time of Christ, it was around 180 million. A thousand years ago it was 265 million. In 1800, it was 900 million and by 1900, it had become 1650 million. Today, it is 7.6 billion, over four and a half times what it was in 1900, just 117 years ago.

What other world-shaping event occurred in 1900? The advent of the automobile. Though its origins occurred in Western Europe, the automobile made a very quick toehold in the U.S. by 1908. That was the year that Henry Ford evolved (from earlier models A through S) the Model T. Ford's aim was to produce a universal car for everyone. It would be so simple and inexpensive that anyone could afford one. It was light, tough, and had high road clearance to negotiate the rutted dirt roads of the day. Ford figured out production techniques that allowed phenomenal quantities to be produced. By the early 1920's, he was producing a thousand per day and half the cars in the U.S. were Model T's. In the early twenties also, the cheapest roadster cost only $260 (6). The Model T's 20 horsepower engine would bring a light car to just shy of 45 miles per hour. The last Model T was built in 1927 and fifteen million had been built. By that time, other makes had become a lot nicer. Ford's business techniques allowed him to survive the Depression that began in 1929. By that time there were dozens of American car manufacturers that perished and relatively few survived.

Numbers of cars operating in the United States
by decade number as follows:

1910: 459,000
1920: 7,500,000
1930: 22,972,745
1940: 27,372,297
1950: 40,190,632
1960: 61,419,948
1970: 88,775,294
1980: 120,743,495
1990: 132,164,330
2000: 225,821,240
2010: 250,070,000
2017: 263,600,000

From 1910 to 2010, the number of cars in the U.S. increased by 545 times.

In the United States there was a great deal of experimentation going on with automobiles regarding size, body type, and motive power- whether it might be steam, electric, or gasoline. Over a hundred makes of 'cycle-cars' were produced from about 1912 to 1914. They were very small and light, usually using a V twin motorcycle engine, friction transmission, and V belt final drive. They sold for $200-$400 and never became popular because, for a slightly greater cost, one might buy a Ford (6).

Other well known manufacturers tried to compete with Ford head-on for cost and desirability in the 1920's. Among them were General Motors with the Chevrolet 490, Durant with the Star, John North Willys with the Overland which, in 1926, became the Whippet. Its four cylinder engine went into the Jeep in World War II. As a rule, all those that tried to compete with Ford made cars a little nicer but never as low-cost. In 1928, the car makes competing with Ford were advancing in sales and Ford, to compete, brought out the Model A. Among its many improvements were four wheel brakes, two-way shock absorbers, twice the horsepower, and a conventional clutch and three speed transmission. The body was totally new and very stylish, resembling a small Lincoln. From 1928 through 1931, right through the Depression, Ford produced 4,858,644 Model A's.

Struggling to compete, General Motors in 1929 came out with a six cylinder Chevrolet of some merit to replace the four. Walter Chrysler, not to be outdone, produced a beautiful four cylinder Plymouth with flexible engine mounts, very nice styling and four wheel hydraulic brakes. Its engine, though only a four, was quite advanced with full pressure lubrication and a hefty counterbalanced crankshaft. Its 56 horsepower enabled it to pull away from a Ford or Chevrolet of the day (7). It is seldom remembered today because Chrysler made only a tenth as many as Ford made Model A's.

After the Depression of 1929 the cars built in the U.S. were large by world standards for four reasons: low cost gasoline, improving highways, moderate ownership fees, and longer trips. In the early 1930's, a person of wealth might wish to impress others by driving a 16 cylinder Cadillac or Marmon. If money was of very little consequence, they might drive a Deusenberg. Its Lycoming straight eight overhead cam supercharged seven liter engine would easily push it past 90 miles per hour in second gear. A friend of the author had an opportunity to drive one from Florida to Detroit and said it felt like he was sitting on top of a runaway freightcar. The poorest fuel economy in the author's experience came from a 1934 Packard V-12 limousine. Around town it got 4 mpg. Even today there are some vehicles of that realm.

Early in the 1900's, Western Europe had a very different set of requirements regarding automotive design. Many of the cities had been laid out years before, with narrow cobblestone streets and very short roads between one another. Gasoline was expensive and the taxes on ownership of an automobile were high. The taxes were based mainly on engine size and vehicle weight. In the early days there was a tiny car called the Austin 7. It was designed to be just big enough for a husband and wife with two small children. Herbert Austin, a former manufacturer of sheep shearing machines, was the maker. The engine was a tiny four cylinder sidevalve of 748cc and, it is believed, the seven referred to taxable horsepower.

Fuel taxes in Europe have always been high and each country produced very fuel efficient cars, most with engines of under a liter (61 cubic inches) displacement. Today many foreign countries produce cars designed for export to the U.S., and these tend to be larger and more powerful, but are still more fuel efficient than their American-built counterparts.

Soichiro Honda bears mention as the developer of (among a great many cars) very polite and reliable motorcycles. One of them was the little Honda Cub. It was a light, quiet, reliable little bike with a 4-stroke engine of .09 liter displacement. Exported all over the world, by 1977 over 21 million had been produced (8).

Generally, it might be noticed that the world's manufacturers of motor vehicles for sale outside the United States strive for high fuel economy. Mentioned earlier in this writing is the fact that in the United States we burn six times as much gasoline per person as in any other country. Thus it might seem that the consumers of the U.S. bear a greater responsibility in solving global warming problems than others that share the earth.

Casting aside those who deny the warming notion, let's take a look at our options in bringing together a cure or at least slow down the growth. The sun is the source of all the energy we have had and will have. It, through the miracle of photosynthesis in green leaves starting about two billion years ago, changed the earth's atmosphere from carbon dioxide and nitrogen to oxygen and nitrogen (1). Today photosynthesis is still occurring in green leaves, but man is beginning to reduce their overall quantity. Forest fires also cause a significant loss of green leaves and increase of CO_2, as well as adding heat to the atmosphere.

Over a great deal of time, plant life rots, releasing natural gas (CH_4) and leaving carbon rich residue like peat, oil, tar, and coal in various combinations. Shale is a fine grained sedimentary rock containing kerogen, an early state of crude oil (9). The U.S. has about 72% of the world's oil shale. Obtaining oil from shale is very expensive and leaves a terrible mess of the earth where it is taken. Before 1900, the earthly reserves of fuel beneath our feet had scarcely been touched. Since that time, the earth's population has grown by 4.5 times and the burning of gasoline has gone from zero to over 320,000 gallons per minute. As we have seen earlier in this writing, all of the heat caused by that burning eventually goes into the earth's atmosphere.

In addition to that heat is the radiant heat from the sun. During the darkness of night, much of this heat radiates into outer space through long wave radiation. The presence of carbon dioxide and methane in our atmosphere hinders the long wave radiation from leaving during the darkness of night. This is the _primary cause_ of global warming. Since the release of methane from deep in the world's oceans is very much increased by a small rise in ocean water temperature, then all phenomena that cause seawater temperature to rise should be minimized. Means of capturing some of the methane that bubbles up in the world's oceans might somehow be developed. However, all the methane we on earth could use would be but a tiny fraction of all there is captured in clathrates in the world's oceans. The foregoing leaves us with a very grim outlook for the future.

The following may illustrate some of our options to improve the outlook to keep the future long. Enhance our educational system by imparting a clearer understanding of our plight.

Bolster this understanding with our new source of collected facts- the internet. Learn critical thinking so that ideas and suggestions that don't hold up in the light of truth can be quickly discarded. Embrace the laws of nature including chemistry, physics, the behavior of heat and light. Make plans that cooperate with those laws. Improve the dispersal of facts about medicine so that people who have no chance of educating their children needn't have those children. Here education means imparting some notion of how the world works.

Several things have been developed within the last forty years that give us a good chance of improving our plight. A major one is the solar panel that takes a tiny fraction of the sun's energy and converts it into instantly useful electric power. Some amount of detail will be presented here because solar panels may become very significant in our future. It has been carefully measured that every square meter (10.75 square feet) of solar collector surface in bright sun receives 1050 watts. The average over a 24 hour days is 164 watts. The energy conversion efficiency of a panel today must be considered as well as how much that efficiency might be improved in the future.

Published on April 6, 2017, in the online edition of Nature Communications is the following: "In theory, 30% is the energy conversion limit for traditional single-junction solar cells as most of the energy that strikes the cell passes through without being absorbed, or becomes heat energy instead. Experiments have been taking place around the world to create various solar cell designs that can lift these limitations on conversion efficiency and reduce the loss of energy. The current world record is 46% for a four junction solar cell." Apparently there is some hope of an energy conversion efficiency of over 50%.

The life span of today's solar panels is 20 years during which time the efficiency degrades about one percent per year. To use solar panels to aid electric cars by charging their batteries while parked bears looking into. The usable surface on top of an average sedan is a rectangle 48" wide by 72" long giving 2.25 square meters. Using panels of 30% efficiency, the output would be 708 watts. One horsepower equals 746 watts so this would charge the car's battery at just under one horsepower equivalent. Not very promising. And this is only in bright sun.

Now suppose there was a twenty car parking lot under a single roof which was covered with solar panels like those mentioned before. Each parking space is nine feet wide and twenty feet long so that now each car can receive the power of sixteen square meters or 5,035 watts, the equivalent of 6.7 horsepower. However, all the electric power conversion units have an efficiency of less than 100% so that the 6.7 horsepower worth of electric power from the solar panels comes out about 4.5 horsepower at the car's wheels- some loss, but still worth considering. At present, huge solar panel arrays are being installed in fields near freeways so that electric vehicles can stop for a charge. Since a full battery charge takes time, a system from bygone days may be adopted wherein a vehicle's discharged battery can simply be traded for a charged one.

To gain a real sense of the sun's radiant energy, walk into an open parking lot on a sunny hot afternoon. Put your hand on top of a white or light colored car. Now put your hand on top of a black or dark colored car. The temperature difference is noticeable. It makes a difference in the work the air conditioner must do and light colored car are more visible in traffic. Some believe this improves safety the same way light clothing does when riding a bicycle. Solar panels on the roofs of homes have proven to reduce monthly electric bills by a significant amount, especially on hot summer days. Here they help in two ways- one by supplying part of the power the air conditioner takes, and two, by shading part of the roof thereby reducing the need for the air conditioner.

At the present time, the internet tells of some very encouraging progress in solar electric power stations. In 2016, the largest photovoltaic power station in the world was the 850 megawatt (million watt) Longyangxia Dam Solar Park, in Gonghe County, Qinghai, China. As of April 2017, India may have the largest solar park with 900 megawatts of the 1000 megawatts already commissioned at the Kurnool Ultra Mega Solar Park.

However, as mentioned earlier, all electric power generated for man's use eventually turns up as heat in the atmosphere. What about a refrigerator, which uses electricity and gets cold? Actually it is a motor driven heat pump that draws the heat from inside and discharges it outside via the condenser, a long serpentine tube, painted black like the hot car roof with numerous wires welded to it to increase its surface in the air. It is usually located below and behind the main box. Feel it when the machine is running. That is the heat the machine took out of the air inside the box.

The main advantage that solar panels have over steam driven power plants is that they create electric power without heat- they don't need steam to make them put out power. Do they take heat from the sun to make electric power? If so, they must actually cool the atmosphere somewhat. The internet gives much discussion on the question, but concise answers escape at this time. However, it is true that all the electric power they produce ends up warming the atmosphere just as it does from any source. It might be mentioned that electric vehicles will ultimately turn the electric power that propels them into heating the atmosphere. This occurs for the same reason that any car converts the energy that propels it into heat in the atmosphere, by churning the air, flexing the tires and the heat losses in the propulsion system.

The heat losses in this system come from the battery charger, the battery being charged, the battery being discharged, the power control system, and the motor and gearing system. The heat losses in the propulsion system of the electric vehicle are far smaller than in those powered by internal combustion engines. The total automotive system with the lowest possible losses is the solar panel array charging the battery of an electric vehicle. Even then it is best if the car is full of passengers and the cruising speed modest.

This automotive system, by all indication, is to be the most energy efficient and have the least effect on global warming that man will ever create. It will not eliminate the warming, but will slow its growth. To put a little more authority into the global warming notion is the following quote from The Intergovernmental Panel on Climate Change Fourth Assessment Report of 2007. "The test in science is whether findings can be replicated using different data and methods. More than two dozen reconstructions, using various statistical methods and combinations of proxy records, have supported the broad consensus shown in the original 1998 hockey stick graph, with variations on how flat the pre-20th century 'shaft' appears." The 2007 IPCC report cited 14 reconstructions, ten of which covered 1,000 years or longer, to support its strengthened conclusion that it was likely that northern hemisphere temperatures during the 20th century were the highest in the past 1,300 years. A 'hockey stick' graph in this case is one where years are plotted horizontally and global warming is plotted vertically. Here the shaft of the hockey stick is nearly horizontal for over a thousand years, then suddenly, like the hitting end of a hockey stick, takes a sharp upturn starting about 1900 and rising almost vertically to 2007 when the report was assembled.

As mentioned earlier, in 1900 there was no gasoline, and the number of people on earth was only 22% of what it is today. Now we are burning three hundred and twenty thousand gallons of gasoline each minute, to propel our 264 million automobiles in the U.S. The time between 1900 and today is but a tiny, tiny fraction of the time man has inhabited the earth. But in that tiny tick of time and despite two world wars, many smaller wars, and world-wide flu and other diseases, we have stuffed ourselves into the world in such numbers that our very atmosphere is threatened.

Since methane is, by far, the worst global warming component of our atmosphere, more detail on its origins is in order. First let's trace the origin of methane to see if all of it came from microbial decomposition of green leafed plants or if there is another source. There are several explanations that say the same thing: the primary source is always the same as stated. The way that methane gets deposited in clathrates (methane hydrates) is from methane production on the sea floor under low temperature and high pressure conditions. These occur predominantly on the continental margins. The warmer the water, the greater the depths must be (higher pressure) to form the hydrate. If the hydrates are either warmed or depressurized they will revert back to water and natural gas.

Following is a broader explanation about our present situation and future plight, focusing on methane, its sources and effects. Since the second half of the 19th century, anthropogenic (human caused) carbon emissions have resulted in greenhouse warming of the globe (Intergovernmental Panel on Climate Change in 2014). This warming has caused considerable environmental impact worldwide. The ice caps have lost 50% of their ice area and 80% of their ice volume since 1950, causing sea level rises of 10-20 centimeters (3.9-7.8 inches)(Berliner 2003). Global precipitation has become more uneven, leading to either drought or flooding conditions. Warmer ocean temperatures and greater concentration of water vapor in the atmosphere are fueling more powerful hurricanes.

Up to 270 ppm before 1880, the atmospheric carbon dioxide concentration is now 404 ppm. (Moss 2010). This is projected to increase at a rate of 2 ppm per year for the foreseeable future, strongly suggesting that the one degree celsius of greenhouse warming that has occured since the industrial revolution will only further increase in the coming decades. This accelerated warming could potentially lead to several significant impacts that could exacerbate warming yet further. Among the most substantial of these impacts is the destabilization of methane hydrates on the ocean floor. These deposits contain vast amounts of methane suggesting that their destabilization could result in substantial additional greenhouse warming should this gas be emitted into the atmosphere. Enough carbon is stored in methane hydrates that if even a small fraction of it were to be released, resulting warming would potentially rival all anthropogenic warming since the industrial revolution. Perhaps what renders methane hydrate destabilization one of the most significant environmental impacts of existing climate change is that initial releases of methane would cause warming that would lead to further hydrate destabilization and methane release. This would cascade in a positive feedback loop that would result in considerable climate warming (Smith 2006). Such global temperature increases could potentially occur very rapidly depending on the rate of venting from these methane deposits, resulting in considerable cost to both the biosphere and human society. While hydrates have been studied since their discovery in the early 19th century, naturally occurring deposits have only been seriously investigated by the scientific community since the 1990's.

That intensive study has begun only recently suggests that such search for accurate answers should expand even faster if we are to gain any reasonable control over our future. Since the burning of gasoline is, at present, the greatest source of global warming, it would seem reasonable that there should be a study tax on it. The study tax would pay for numerous organizations worldwide which at present are approaching agreement on how gasoline burning fouls the world's air and exacerbates other atmospheric problems. At present, gasoline tax in the U.S. is about half what it is in Western Europe. Large powerful cars and huge pickup trucks abound in the U.S., all to the profit of their makers and the oil companies. Something about young men driving huge pickups may be akin to those who, in 1917, flew fighter planes. Typically they might drink blackberry brandy before takeoff to dull their fear and tighten their bowels against the haze of castor oil fumes flying from their rotary engine. Castor oil was used to lubricate these engines on a constant loss system leaving partly burned oil landing on the pilots. Being shot at with intent to kill or enflame may have only heightened the thrill. Then, too, if you get killed, you become a hero and, in some cases, avoid the rigors of finding yourself.

The study tax could improve our understanding of methane gas bubbling up in the world's oceans and its ultimate effect on the liveability of our planet. It might hopefully change the minds of doubters who find peace of mind in joining the ranks of those who say it isn't so. Many reputable groups may be found who study and measure and find out what has been, what there is now, and how fast it is changing. This in regard to population, burning of gasoline, CO_2 concentration, methane release rate from the world's oceans and other parameters that will affect man's future on earth. To the author's way of thinking, coming to understand what is- even if undesirable- beats hoping that it isn't while not understanding. Then the direction to direct our efforts is clear.

References

The Way Nature Works, many contributors, MacMillan Publishing Co.

The Lore of the Train, C. Hamilton Ellis, Madison Square Press, Grosset and Dunlap Inc.

Internal Fire, C. Lyle Cmmins, 1976 Carnot Press, reprinted 1989 SAE Inc.

Great Inventions and Discovered, Donald Clark, Marshall Cavendish Books, Ltd.

The Steam Turbine by R.H. Parsons; the British Council by Longmans Green and Co.

Standard Catalog of American Cars by Beverly Rae Kimes and Henry Austin Clark, Krause Publications

Chrysler Engines 1922-1998 by Willem L. Weertman, SAE International

Honda; the Man and his Machines by Sol Sanders, Little, Brown, and Company

Oil 101 by Morgan Downey, 2009

ABOOKS

ALIVE Book Publishing and ALIVE Publishing Group
are imprints of Advanced Publishing LLC,
3200 A Danville Blvd., Suite 204, Alamo, California 94507

Telephone: 925.837.7303
alivebookpublishing.com

www.ingramcontent.com/pod-product-compliance
Lightning Source LLC
Chambersburg PA
CBHW020210200326
41521CB00005BA/323